# Earth Sheltering:
## The Form of Energy and
## The Energy of Form

## Pergamon Titles of Related Interest

**Bergman** Rockstore 77

**Bergman** Subsurface Space (Rockstore 80)

**Hoek** KWIC Index of Rock Mechanics Literature
1870-1968

**Holthusen** The Potential of Earth-Sheltered and
Underground: Today's Resource for Tomorrow's
Space and Energy Viability

**Jenkins** Kwic Index of Rock Mechanics Literature
1969-77

## Related Journals*

Advances in Tunneling Technology and
Subsurface Use
Futurics
Geothermics
Technology in Society
Transportation Research
Underground Space

*Free specimen copies available upon request.

# Earth Sheltering:
# The Form of Energy and
# The Energy of Form

Editor: T. Lance Holthusen
Author: Edward R. Frenette, AIA

For: The American Underground-Space
Association, in conjunction with their
1981 Design Competition

Cover Graphic: Robert Hull
Cover Design: Katherine Reif Anderson

PERGAMON PRESS
New York/Oxford/Toronto/Paris/Frankfurt/Sydney

Pergamon Press Offices:

| | |
|---|---|
| U.S.A. | Pergamon Press Inc., Maxwell House, Fairview Park, Elmsford, New York 10523, U.S.A. |
| U.K. | Pergamon Press Ltd., Headington Hill Hall, Oxford OX3 0BW, England |
| Canada | Pergamon Press Canada Ltd., Suite 104, 150 Consumers Road, Willowdale, Ontario M2J 1P9, Canada |
| Australia | Pergamon Press (Aust.) Pty. Ltd., P.O. Box 544, Potts Point, NSW 2011, Australia |
| France | Pergamon Press SARL, 24 rue des Ecoles, 75240 Paris, Cedex 05, France |
| Federal Republic of Germany | Pergamon Press GmbH, Hammerweg 6, Postfach 1305, 6242 Kronberg/Taunus, Federal Republic of Germany |

# Acknowledgements

This volume includes the excellent work of fifty designers who participated with over 100 others in the 1981 American Underground-Space Association's Design Competition; the first such competition in America to focus on a full spectrum of earth sheltered and underground works-in-progress. On behalf of the AUA I would like to recognize all of those who entered as being the kind of risk takers that offer their time and talent to move an important genre of research and design ahead. We think that the designs and narratives contained herein will demonstrate how rapidly and thoughtfully this energy and space-saving field has evolved in a very short period of time.

The AUA wants to express a very special thank you to all of the people who helped create this volume. Our publisher, Pergamon Press, and especially Robert Maxwell, Publisher, were instrumental in making the Design Competition happen. We wish to particularly recognize the generosity and skills of Richard Vasatka, President, and Ed Frenette, Director of Design, Setter, Leach & Lindstrom, Inc., a Minneapolis based architecture and engineering firm which has completed numerous underground and earth sheltered projects. Ed Frenette has contributed countless hours to the competition and book through his many roles as technical consultant and jury referee for the competition and as author of several sections of this book. The Design Competition jury composed of Edward Allen, Thomas Bligh, Antonio DiMambro and Chester Sprague, all faculty at the Massachusetts Institute of Technology, have our deepest gratitude for the time and expertise they so willingly applied to what proved to be a very challenging task.

We also wish to particularly thank our technical staff for their patient cooperation in producing this volume. They include: Kathy Reif Anderson, Up North Design and Mary Rollwagen, TLH Associates, Inc., for cover design, layout assistance and proof reading; Theda Stearns, TLH Associates, Inc., for her assistance in organizing and coding the competition; the secretarial staffs of both TLH Associates, Inc. and Setter, Leach & Lindstrom, Inc.; and Robert Bosacker, Roberts Litho, our printer.

I would also like to take the liberty on behalf of all those who participated in this competition to thank the American Underground-Space Association for sponsoring this competition in conjunction with the Underground Space Conference and Exposition held in Kansas City, Missouri, June 8-10, 1981, the first of what we hope will be a regular activity. Thomas Atchison, Executive Director of the AUA, should especially be singled out for his ongoing encouragement for all activities of this kind which have contributed so importantly to the growth of this field.

Once again, the real debt is to all of the architects and engineers who invested their late nights at the drawing boards producing the designs contained herein and to Ed Frenette for his continued guidance.

T. Lance Holthusen, Editor
St. Paul, 1981

# Contents

# Contents Continued

# Contents Continued

# About the Author and Editor

## Edward R. Frenette, AIA

Ed Frenette is Director of Design for Setter, Leach & Lindstrom, Inc., a 90 person architectural and engineering firm located in Minneapolis, Minnesota. He received his Masters of Architecture from the University of Toronto in 1977. He was a Guest Critic at Pratt Institute and Columbia University in New York City, 1969; a Teaching Assistant, University of Toronto, 1976-77, a Visiting Critic at the University of Minnesota 1978-81. In addition, he chaired the Michigan Masonry Design Awards Program in 1979 and the Minnesota Society Honor Awards Program, 1976-79. Ed is currently a member of the *ARCHITECTURE MINNESOTA* Editorial Board.

Ed Frenette is primarily interested in architecture as a medium for the expression of Social Values, Urbanism, Regionalism, Energy, and History. His lectures on the subject have included: "Understanding Architectural Media" at the Institute of Culture and Technology, Toronto, 1977; "The Generalist Vs. the Specialist," Conversations with Romaldo Giurgola, Marshall McLuhan, Cesar Pelli and Malcolm Wells, University of Toronto, 1976-77; "Beyond Theory and Perfection in Architecture," University of Minnesota, 1979; "Humanizing the Institution," University of Minnesota, 1980; "The Form of Energy — The Energy of Form," North Dakota State University, 1980; and "The Least Post-Modern," M.I.T., 1981.

His published articles include: "The Environmental Learning Center," *Underground Space,* London, 1979; "Form and Energy," *Architecture Minnesota,* 1981; and "The Form of Energy — The Energy of Form," Pergamon Press, Inc., 1981.

Ed received a National Endowment for the Arts Award to support his urban design work in New York's East Harlem in 1969; a University of Toronto scholarship to support his urban design thesis in 1977; an Urban Design Award for the Duluth Clinic, Duluth, Minnesota, 1979; an MSAIA Special Award in Urban Design for the Public Gallery addition to The Science Museum of Minnesota, 1979; a Merit Award for his design of an Energy-Efficient Suburban Community, 1980; and this year his drawings on "Urban Design" are being exhibited at The Walker Art Center, Minneapolis, Minnesota; Chicago Public Library Cultural Center, Chicago, Illinois; Blaffer Gallery, Houston, Texas; Neuberger Museum, Purchase, New York. His constructed works include over 50 buildings in Minnesota and Wisconsin.

## T. Lance Holthusen

T. Lance Holthusen is President of TLH Associates, Inc., a St. Paul, Minnesota based consulting firm which assists organizations in preparing for and shaping their future. He received his Bachelors in Political Science from the University of Wisconsin-Stevens Point, and his Bachelors and Masters of Divinity from Luther Theological Seminary in St. Paul, Minnesota. Further graduate studies have focused on alternative futures education. Mr. Holthusen has been a visiting lecturer and adjunct faculty for St. Cloud State University in the Departments of Education and Industry and Technology. He was formerly a consultant to The Science Museum of Minnesota and originator and Director of its Futures Studies Department.

He is primarily interested in corporate and public understanding of alternative futures specifically in the basic life support systems of food, health, energy, education, shelter and natural and man-made environments. He is a past officer and life member of the World Future Society, Past-President and ongoing Director of the Minnesota Futurists-WFS, a member of the World Future Studies Federation, Rome, Italy, and Managing Editor of *Futurics,* Pergamon Press, London and New York.

# Earth Sheltering: The Form of Energy and The Energy of Form

by
Edward R. Frenette, AIA
Director of Design
Setter, Leach & Lindstrom, Inc.

## The Energy of Form

Significant leaps in the continuum of architectural development have often been ushered in by timely competitions or their built equivalents, expositions.

Aspiring modernists saw the 1893 Columbian Exposition at Chicago and the decadent eclecticism it spawned as a great leap backwards. The League of Nations Competition of 1927 crystallized the positions of both modernist and classicist. In the 1922 Chicago Tribune Tower Competition the modern designs that lost were at least as influential as the classical designs that won. More recently, The Architects Collaborative's sleek Johns-Manville Headquarters outside Denver, Michael Graves' Portland Government Building and Romaldo Giurgola's Parliament House for Canberra, Australia have clearly articulated the state of current modernist, post-modernist and eclectic thinking.

The collective appearance of new architectural forms has had a powerful and energizing effect on architectural development. The causal relationship, however, is not clear. It is entirely possible that changing ideas first clearly manifest themselves in competitions rather than that competitions establish new direction. Evidence indicates that either or both can occur and often do.

In view of this potential for a competition to be both a method of sampling the state of the art and a catalyst for idea development, the American Underground-Space Association (AUA) decided to sponsor a national competition to find "new directions in earth-sheltered construction."

The impetus for this effort came from those within the AUA membership who believe that earth-sheltered design is at an awkward, adolescent stage. Adolescent because knowledge about heat transfer, life-cycle costs, and psychological effects of earth-sheltered design are only now becoming well known after a decade of maturation.

Awkward, because it is not yet fully understood how the specifics of earth-sheltering should be considered in context with other emerging technologies such as passive solar and super insulation, or how all should be considered within the more general and mature disciplines of urban design, historic preservation, planning and architectural theory.

## Work in Progress

The AUA solicited entries of "Work in Progress" by both students and professionals in four categories — single family residential, multi-family residential, non-residential and research. The "Work in Progress" theme was an attempt to stimulate active participation. Special projects, which can cost a firm upwards from $2,000 to prepare, would clearly not be required. Because most projects would presumably be responses to real clients, budgets and sites, the competition would also be an accurate indicator of the state of the art. Three awards were offered in each of the four categories — first, second, and merit. In addition, student and professional work was judged separately, making 24 awards, totaling $11,200, possible. Individual awards range from $200 to $1,000.

Residential design was separated into single-family and multi-family categories because, it was felt, the difficulty of doing successful multi-family housing of any type, let alone earth-sheltered, might unduly bias the entries and awards in favor of free-standing small houses, the traditional playground and laboratory of the architectural avant-garde. Though awkward, non-residential is still the best term for describing commercial, institutional and industrial projects that have no more specific nomenclature. Research was intended in its most general definition to be anything which would add to the understanding and development of earth-sheltered construction without leading directly to a built project.

## Judging the Pragmatism and the Poetry

The jury was comprised of four members of the MIT facility who both teach and consult professionally on architectural or earth-sheltered projects.

Edward Allen, chairman of the jury, is an associate professor in architecture at MIT. He specializes in teaching architectural design and technology. His professional practice includes designing numerous solar heated projects and his published works primarily concern design through the use of appropriate technology.

Chester Sprague, an associate professor of architecture at MIT, also teaches design with concentration on housing design and community planning for severe climates such as Alaska and the southwest. In addition, his professional practice includes design and study of educational and work places in both the U.S. and England.

Antonio Di Mambro, an assistant professor in Architecture at MIT, has a broad background in urban design and planning. His professional practice includes the design and planning of large scale projects in the U.S. and Italy.

Well known to earth-sheltering devotees, Thomas Bligh, a professor of mechanical engineering at MIT, was the only non-architect on the jury. Dr. Bligh qualifies as one of the founding fathers of earth-sheltered design due to his early and extensive research and writings on heat transfer in below-grade buildings, solar heating and cooling, and numerous other related subjects. As a fringe benefit, Bligh brought a very keen eye for aestheticcs to the jury thanks to his experience as a consultant on many innovative energy conserving projects.

Like all good juries, this one took the task of judging over 170 projects submitted by students and professionals from all across the country very seriously. First, criteria were established governing consideration of the methods of energy conservation and their effects on the architectural result. Conservation technology and architecture were each considered on their own merits. Although some projects exhibited their technology more dramatically than others, quality spaces created as a result of the integration of technology with traditional architectural and cultural values were prioritized over technological exhibitionism.

Again, like all confident juries, they listened to the previously developed guidelines for the competition and then augmented them with their own. They elected to give only 14 awards of a possible 24, and only three first prizes of a possible eight. In addition, they created a new category, Honorable Mention, which did not receive a cash prize. Nine awards were made in this new category. Because prize money was not awarded, more example projects could be included here. To the credit of the jury, the pattern of awards given and not given accurately indicates the varying degrees of difficulty in applying earth-sheltered design to the different building types and the varying quality of the resulting entries.

The pattern of awards is indicated by the following figure which clearly shows the few stars, 30% First and Second awards, the many above average, 70% Merit & Mention awards, and the problem areas, multi-family residential and research.

| CATEGORY | First | Second | Merit | Mention |
|---|---|---|---|---|
| *Single-Family Residential* | | | | |
| Professional | * | | * | ** |
| Student | | | | *** |
| *Multi-Family Residential* | | | | |
| Professional | | | ** | |
| Student | | * | | ** |
| *Non-Residential/ Commercial* | | | | |
| Professional | | | **** | |
| Student | * | * | | ** |
| *Research* | | | | |
| Professional | * | * | | |
| Student | | * | | |

# Criticism for the Non-Critical

The criticism of architecture is a much debated issue. The inclusion of earth-sheltering in the debate only complicates the equation. Consequently, the jury deliberately tried to evaluate projects according to separate and distinct criteria first, and then judge them as a whole.

Since architecture exists to be inhabited, it was first judged on its own ground before its below grade merits were considered. Once projects of some architectural merit were identified, all were reviewed again to determine those which fully utilized the concept of earth-sheltering. At this point, two categories of earth-sheltered projects emerged — those placed below grade primarily to conserve energy and those placed below grade for many reasons, one of which may be energy.

It was in the linking of the evaluations of architectural and earth-sheltered merits that the jury had its most spirited debate — not in regard to the best projects but in regard to those considered borderline. In understanding how the jury ultimately linked these two issues, it is helpful to understand the results of the research on earth-sheltering now being completed for the Navy Facilities Engineering Command. (Summary has been included at the end of this introduction.)

A vulgar yet accurate rule of thumb can be deduced from this new research. It is that the methods used and the costs incurred to accommodate earth-sheltering should be modest if the prime impetus is economic. The savings of energy must be carefully analyzed relative to the cost of construction. In addition, earth-sheltering should be considered in concert with other conservation technology such as super insulation and passive solar design to determine which individual or collective techniques can best be employed. The designers of the best projects seemed to understand intuitively what is only now being verified empirically.

# The Form of Energy

Many who have reviewed the published projects have been amazed at the innovative brilliance exhibited. Robert Hull's house for Washington state, Scott Pozzi's church for San Francisco and Alfredo Devido's house design all express the poetic potential of the structural necessity of earth-sheltered design. All these designs transfer excessive roof loads to the ground economically without inhibiting their spaces. The spatial and structural concepts are in such close synchronization that one cannot conceive of the house without the tree-like columns or the church without its dome. In addition, Hull's house utilizes rather innovative methods of separating the direct solar gain area of the structure from the spaces which must maintain a more moderate temperature. His use of circulation space as the linkage between spaces which are constantly in use and must be comfortable, and those which can afford radical swings in temperature is a positive step away from the ubiquitous greenhouses which pockmark many energy conscious designs. Even the garage door supports used in his projects are handled more like the elements of a desirable pergola than the required solar control devices they are.

It is interesting to note that most philosophies of architectural thought are represented here. Leading off are Robert Hull's two houses which can be termed late-modern for their geometric clarity and technologic expressiveness. Following are two houses which are most definitely post-modern. Gordon Ashworth's design for Manhattan, Kansas, is particularly notable for the way it utilizes the potentials of earth-sheltering to create interior spaces and to accommodate difficult site constraints. It has prototypical value in that it demonstrates how a new earth-sheltered structure can be sensitively sited among more traditional structures. A pile of sod decorated with solar collectors can be particularly offensive juxtaposed to a well mannered tudor or Italianate neighborhood.

Another prototypical quality of Ashworth's design, also used by Milo Thompson in his house for northern Minnesota, is the manner in which the submerged surfaces are manipulated. Surprisingly, few architects have taken advantage of the fact that non-exposed walls can take any shape desired in order to affect heat transfer, to create appropriate structural shapes or to shape interior spaces without affecting the visible exterior. Complex submerged shapes are also more easily insulated than their exposed counterparts because insulation is applied over a submerged wall rather than contained within an exposed one.

Both Ashworth and Thompson used the earth-covered walls of their modest houses as an architect of the baroque tradition used stone to create the poché between spaces of conflicting geometries in their immodest cathedrals. Like Ashworth's, Thompson's design is interesting for its response to context: in this case, forest primeval rather than urban.

Both Robert Banbury and Alfredo DeVido, in their designs for Ohio and New York residences, have found convincing places to locate solar panels that can't always be part of the structure. Although such locations on adjacent berms raise questions about maintenance and vandalism, they avoid the irony of locating humans below grade with machinery on top.

Few projects dealt directly with natural ventilation and none as poetically as Gunnar Birkerts' residence for Michigan. His design is convincing in the way it utilizes passive solar, natural ventilation and earth-sheltering in a form which metaphorically relates to both the gentle rolling hill it's set within and to the lazy movement of August breezes it attempts to channel.

Three residential typologies emerged as common methods of organizing sublevel spaces and directing them towards the sun. One plan is spine-like with corridors and rooms attached like leaves to a stem, such as Hull's (circulation located at the front), Thompson's (circulation located in the middle) or James Tune's (circulation cleverly located along the rear). Also very common was the single facade approach used in the design by O'Neill Perez & Associates. Less common was the row-house-like concept used by Birkerts and DeVido.

## Housing Below Grade

The awards not given for multi-family housing tell as much about this category as the winning entries do. A few years ago, competition juries were quoted as saying "Single-family housing is not a valid test of design ability. The real challenge is in housing not houses."

Judging from the entries, the profession is not yet up to the challenge. The lack of good entries in this category suggest two conclusions. Solutions to the basic problems in multi-family housing — the relationship of public and private space — are of such a priority that earth-sheltering should be employed only when particularly appropriate. This tactic would avoid adding earth-sheltered housing to the list of modern abandoned and demolished projects that now litter our urban areas and modernist dogma. Another conclusion is that future competitions should concentrate on this category, calling out the mixed priorities inherent in the task in an effort to stimulate invention and understanding.

In light of the jury's awareness of the difficulty of the task, it was particularly rewarding for them to find a design as well done as Milo Thompson's housing project for a reclaimed gravel pit in Minneapolis, Minnesota. The site plan should be of interest to anyone contemplating any kind of housing. The concept of "nesting" one unit over the other which allows bi-level access from grade and the use of earth berms provide great privacy to a development which would otherwise seem very dense. Allen Brown's student project has many of the attributes of Thompson's but his unit plans are perhaps better resolved. Through the use of earth-sheltering, both projects appear virtually half as dense and twice as private as they are actually — no mean trick when compared to the modest advancements that are normally made in this building type.

David Gibson's winning design for Aspen, Colorado, raised more questions about the uses of earth-sheltering in housing design than it answered. The design places a motel-like workers' housing project beneath a park where it does not offend the land owners adjacent. The jury decided to give the project the benefit of their doubts. They considered it a short-term residence for people who are acquainted and socialize with each other. This rationale resolved the social, political and privacy questions raised by the design and allowed awarding a prize to a skillful site plan and the further discussion of the issues.

## From Desert Object to Urban Fragment

Projects have been organized according to similar attributes to allow readers to learn as much from the range of the projects exhibited as from the individual examples. This approach is perhaps more evident in the commercial chapter where projects are arranged from free-standing and autonomous to urban and connected.

Although it is free-standing, the Visitor Center by the Colyer/Freeman Group of San Francisco is certainly connected visually to its site. The form could have been more fluid but use of alternating rusticated and smooth masonry bands as a metaphor for the area's geology is a simple and effective method of relating the building to its site. This same pragmatic approach is evident in the simplicity of the cooling tube used. However, it is not clear whether the tube is long enough to allow the earth to cool the ventilation air and not so short that the air eventually heats the earth.

Another political solution to a problem of siting functions where the neighborhood does not want them is demonstrated by the confident design of a convention center for San Francisco by Hellmuth, Obata and Kassabaum. There is no pretense of an economic or energy base to the design. Instead, earth-sheltering is used to solve the important urban design problems of one of the nation's best loved cities. Skillful and sensitive designs such as this for a building type which can seldom be termed sensitive should be encouraged. Perhaps it is not too late to retrofit the Transamerica Tower with earth-berms.

Two prototypical urban-design uses of earth-sheltering emerged during the review of entries — the siting of a project considered undesirable by neighbors and the modern extension of a historic building. DAT Consultants' design of an Italian-American cultural center for New York is a particularly poetic example of the latter. Here the designers have the self-confidence to create a rather passive addition which allows the existing house to stand out as the focal point of this site.

The research projects speak for themselves. Suffice it to say that the only task was to separate the pseudo research — sophomoric architectural speculation — from the real research. The remainder was thoroughly excellent and all won awards.

## Savings at Any Cost

Not everything came up roses from the earth-sheltered entries. There was definite confusion as to the appropriate uses of the technique for energy conservation. This confusion led to the ironic employment of huge concrete frames for the support of titanic amounts of earth over modest houses and offices. Although the research

referenced earlier does not pretend to be the definitive answer, most who understand the relationship of construction cost to the potentials of energy savings intuitively understand that true economy can only be achieved through balanced use of energy saving techniques appropriate to the specific situation. This normally means modest and sophisticated use of earth-sheltering in designs, not dynamic exhibitions.

The same confusion caused designers to build in rock storage areas or to build up small mounds of earth in the air at enormous material — and often visual — cost in an effort to store the heat energy. The irony is that the heat sinks so dearly paid for in such schemes are available free around every earth-sheltered building by definition. Again, if energy and economy is the name of the game then the earth around an earth-sheltered structure must be utilized fully for its storage and moderating potential. If aesthetics is the name of the game, perhaps energy conservation must be discussed in context with other forces which are now changing architecture.

## Earth-Shelter in Context

The answer to the question, "What will influence architecture tomorrow?" depends on whom you ask and when you ask it. In the rush to house our returning veterans in the 1950s, our elderly in the 1960s and the baby boom in the '70s, few questioned the logic of the elevator highrise and superblock development. It seemed imperative that we replace the old with urban renewal. Fewer still predicted the importance of energy in the early 1970s and virtually no one anticipated earth-sheltered and passive solar design in the early 1960s.

Now, however, history is precious. Neighborhoods and downtowns seem to hold the answers to problems created by towers surrounded by parking lots — and, oh yes, no one disputes the value of energy. Americans have new questions about urbanism, historicism and energy, and they look to professionals for expert answers. The traditional response has been single-issue, specialized answers. Planners look after urbanism, historians preserve our past, and some designers specialize in earth-sheltered and passive solar design. Each expert, of course, predicts a future shaped by the application of his or her expertise. As in the past, however, the future will be shaped from all of society's concerns. This is a comparative look at different responses to energy and a glance at urbanism, historicism and symbolism to see how energy fits in with the other issues that are now forming our future.

## The Least Post-Modern

The growing concern for energy conservation is by no means the only force changing the form of architecture nationally. During the last two decades, society's dissatisfaction with the orthodox-modern response to rebuilding our cities, reusing our historic buildings and using our natural resources has manifested a dramatic change in architectural practice. The modern response to our cities was urban renewal with single-use towers set within single-use superblocks. Once the theory was built, however, many found the towers foreign and uninviting. Everyday needs and desires — like employment, shopping and recreation — were separated by long car rides, or put another way, only available to those who owned cars. Our new crop of affluent city dwellers shows signs of becoming bored with the obligatory fountain-filled plazas. Many long for the messy diversity of older traditional districts such as Georgetown in Washington, D.C.

When most of our building stock had completed its first life in the 1960s, the modern response was to replace or bulldoze. When our most cherished landmarks were threatened, however, architects joined ranks with the historical societies and others to call a halt to the destruction.

Unfortunately all of modernism's energy-responsive models were designed for hot, arid or moderate climates. What worked reasonably well in Germany, France, England and India had to be heated and air conditioned in Minneapolis and Boston. Consequently, when the 1973 Arab oil embargo slowed energy importation, Americans were caught with furnaces and air-conditioning units heating and cooling glass boxes with fixed windows in a climate more severe than most, in a region with dwindling or changing natural energy resources. Unlike the countries that have embraced the forms of the modern movement with climatic impunity, America's severe and diverse climate makes the same forms less adaptable.

A new theory for dealing with our cities, housing and history is "post-modernism," meant literally to be that which has come after modernism. Post-modernism's urbanism is street-related, mixed-use rather than isolated and single-use. Its housing is low-rise high-density rather than highrise. It deals with history through preservation, restoration and adaptive reuse rather than with refacing or erasing. These three approaches have proven to be mutually supportive. Together they have generated interest in the less pragmatically important shortcomings of the modern movement

— symbolism and decoration. Dissatisfaction spawned a new theory, which is now being formalized and put into practice.

Ironically, approaches to energy responsiveness have not been formalized. That is to say, there is no real understanding of what we mean when we say a building is "energy efficient." There is no consensus between the public and the profession about the form of energy concern. There is little doubt, however, that the public and the professional alike desire to manifest their concern for energy in built form. The magnitude of this desire gives rise to the myriad of advertisements, conferences, articles and books now flooding the popular and professional media.

In addition, there is a definite schism between those professionals primarily concerned with urbanism, historicism and symbolism (post-modern concerns for lack of a better term) and those concerned primarily with energy conservation. The former feel uncomfortable with visually polluting their creations or preservation efforts with the hardware of energy conservation. Conversely, the energy camp considers an overconcern with architectural form a somewhat dubious attitude in comparison with the righteous need to conserve our nation's resources. Both positions can be a bit self-righteous. The results are structures that misinterpret or narrowly edit this epoch's wide range of concerns. Both positions can be informed by the diversity resulting from this competition.

## From Prototype to Archetype

The profession has traditionally abhorred categorization according to visual qualities. Few want to be known as glass-box, post-modern or even earth-sheltered architects. Professionals wish to avoid being associated with a cliche while clients enjoy the clarity of simple classification and search for archetypes.

Energy efficiency is the least post-modern of those concerns that have broken down the modern movement. Advocates of energy conservation see it as a technological advancement and economic necessity while others see technology being expressed and prioritized over cultural concerns. However, the value of lining up the following group of buildings according to perceivable qualities — dare I say visual — is that we can then create a visual language — a set of shared ideas about the meaning of architectural form. With that language, we can communicate and compare what is a cliche and what is an archetype.

## Our Formal Future

Unlike the earlier treatment of the subject, this discussion is an attempt to look at the forms generated rather than the technically novel ways energy can be conserved. It assumes that energy can now be conserved in many ways and that we should begin to look critically at the architectural judgments. Architecture has been emphasized, the ubiquitous energy diagrams have been avoided.

The work "in progress" illustrates the range of formal developments possible on a given project. The energy-responsive museum of Brown, Healey & Bock demonstrates how modern forms can be inserted into a quiet setting while preserving the semblance of urban and historical continuity as well as conserving energy. In contrast, the "hightech" functional and energy-conserving techniques of the residence by Robert Hull are given free autonomous modern expression. In one case, aggressive modern forms are played down by earth cover to allow them to sit solidly with their masonry neighbors. In the other case, modern forms are played up and stand lightly on the plain.

The potential of collaging urban, historical and energy concerns can be seen in the residence by Gordon Ashworth. Here modernism is the formal starting point, but its language is extended to include a concern for urban context as well as expression of earth-sheltered techniques.

The extremes of the formal possibilities are illustrated by the Italian-American Cultural Center and the Vacation House by Milo Thompson. No visual expression of energy conservation is evident in the center. Instead continuity with the existing structure and its neighborhood is the highest priority. In sharp contrast, the Vacation House is radically new, yet its newness is instantly "comfortable." Here perhaps is a new vernacular, if those terms can be used together.

## Energetic Pluralism

One could conclude that no one typology is represented here but rather parts of five: "modernism submerged," "modernism directed," "hybrid forms," "new expressionism" and "no change."

Another conclusion is that most designers concerned primarily with energy conservation start with modernism and submerge it for earth-shelter or direct it towards the sun for solar applications. Such a tendency of starting with the well-known base of orthodox modern forms and modifying them experimentally is analogous to how most of the other (post-modern) new design directions have been developed. However, there is an inherent danger in over-concentration on energy in design. It can result in the stunted development of general architectural content. For example, many of the designs appear as free-standing autonomous art objects devoid of connections to their physical, cultural or historical context. Connections with earth berms are often ambivalent and forms seem to be exploding from the natural context more often than they appear to rest within it.

More ironically, over-specialization can result in energy being conserved in a narrow sense — heat and cooling energy — while it is squandered in a holistic sense — on construction or maintenance cost — as discovered during the earth-sheltered study appended here. Another conclusion, which can be drawn from the architectural diversity of the illustrations, is that the concern for energy conservation can either encompass or be subsumed by all other design directions prevalent today or anytime in the future. Such a direction is certainly liberating for both the profession and our clients. It means that each situation can evoke its own response and that our future will be as diverse as our past.

## Synthetic Tradition

Still more exciting is the thought that the 1980s may product an architecture that synthesizes the diverse issues that fragmented so much of our work in the 1970s. It is often argued that contemporary architecture (Modernism, Postmodernism, Late Modernism or the Nextism) has lost its public meaning because society has lost its traditions. Perhaps we can put that meaning back by manifesting the consistent, if not traditional, societal issues of the past 20 years in our architecture. If the connection is understood, it will be our tradition. If not, it will be synthetic, and that too, one can argue, is our tradition.

## Editing the Editorials

The editor's intent has been to select projects which are representative of trends observed, and not only to exhibit good examples. Many projects are included specifically to stimulate broader thinking about the topic. At what point do provisions for saving energy cease to be logical and economical? When do such provisions become only expensive energy decoration? Can the cost in dollars and space of such symbolism inhibit the resolution of traditional urban design and architectural issues?

The intent of editing the graphics and narratives supplied by each designer has been to keep as much of the individual style as possible even when the editor disagreed on the content. However, for the sake of clarity, it is important to know that the most edited statements were those about the "fantastic insulating value of earth cover." In this case, insulating was changed to moderating since all my physics professors agree that earth is a moderator and is not an insulator.

It is comforting, for one who values the ability to articulate ideas, to know that there is a direct correlation between the clarity of the narratives and the quality of the architectural product. Unfortunately, the correlation is somewhat obscured for the reader by the editing that was necessary.

It may be discomforting to learn this late in the introduction that the reader has been given a problem to solve rather than a solution. The problem is to look critically at each project, whether they have a pedigree of an award or not, to determine their value. The real solution is to learn to look critically at all solutions, lest they become problems. The rewards of such a self critical approach are the many brilliant projects contained here that go beyond the specific concern of earth-sheltering to become great architecture on any grounds.

# Appendix

*Earth-Sheltered Construction* is the title of a textbook which is now being prepared by Setter, Leach & Lindstrom, Architects, Engineers, Planners and the University of Minnesota Underground Space Center, Minneapolis, Minnesota. The intent of the text is to provide designers with the comprehensive information and graphic standards necessary for the planning, design and construction of earth-sheltered structures.

The majority of the text will be based on related topics, previously not assembled comprehensively, such as passive solar design, or building codes, edited to make them applicable to earth-sheltered construction. Strategic chapters of the text, such as predesign decisions, energy analysis and life cycle costs, will be based on original research now being completed.

This research has involved computer analysis of energy use and life-cycle cost of three building types: high internal heat load, normal internal heat load and minimum heating; three configurations: multistory, single-story and large single-story; four levels of earth-sheltering: exposed, half-berm, full-berm and bermed-and-covered; three passive solar orientations: south, east/west and north; two passive solar configurations: shaded and unshaded; and six climates: Minneapolis, Boston, Washington, Jacksonville, San Diego and Manila.

As expected, preliminary findings indicate that the benefits of earth-sheltering increase with the need for heating energy and with the severity of the climate. Surprisingly, bermed structures can be less expensive than exposed structures in northern climates due to the savings in footing design and exterior wall materials. However, completely earth-covered structures with large roof areas have not proven to be cost-effective. Earth cover on structures with moderate roof areas, whether single-story or as many as three stories, generally are cost-effective in climates with significant heating loads. In addition, infiltration control, air exchanges, humidity and maintenance have proven to be more significant factors in earth-sheltered design than originally anticipated.

The original research and textbook was commissioned by the Navy Facilities Engineering Command for non-residential, non-defense construction.

# Bibliography

## Directions in Architectural Design

Kenneth Frampton, *Modern Architecture, A Critical History,* New York: Oxford University Press, 1980.

Robert A.M. Stern, *New Directions In American Architecture,* New York: George Braziller, 1968.

Charles Jencks, *Late-Modern Architecture,* New York: Rizzoli, 1980.

Charles Jencks, *Post-Modern Architecture,* New York: Rizzoli, 1977.

Paul D. Spreiregen, FAIA, *Design Competitions,* New York: McGraw-Hill, 1979.

## Directions in Energy Responsive Design

Victor Olgyay, *Design With Climate,* Princeton University Press, 1963.

The Underground Space Center, University of Minnesota, *Earth-Sheltered Housing Design,* Van Nostrand Reinhold, 1978.

Ed Frenette, "Energy and Form, Minnesota's Two Design Imperatives," *Architecture Minnesota,* May, 1981.

# About the Jury and an Interview with Edward Allen

## Edward Allen

Edward Allen is Senior Partner in the architectural firm of Edward Allen and Douglas Mahone, and is Associate Professor of Architecture at MIT. He holds a Bachelor's Degree in Architecture from the University of Minnesota (1962), and in 1964 he received a Masters in Architecture from the University of California. From 1966 to 1968 he studied as a Fulbright Scholar in Italy. The majority of Allen's projects as an architect are solar designs. In addition to numerous articles in both professional and popular journals, he has published *Stone Shelters* (MIT Press, 1969), *The Responsive House* (Editor, MIT Press, 1974), *Teach Yourself to Build* (with Gale Beth Goldberg, MIT Press, 1979) and *How Buildings Work* (Oxford University Press, 1980).

## Chester Sprague

Chester Sprague is an architect and an Associate Professor of Architecture at MIT. He received a Bachelors in Architecture from the University of Minnesota in 1954 and a Masters in Architecture from MIT in 1958. He is currently teaching design and directing a group of projects studying tractability and transformations in housing and neighborhood forms. His past work has included a study of housing conditions in Alaska (report: *Housing in Village Alaska — Background and Alternatives*), a study of cultural facilities program and community participation (report: *Cultural Facilities in American Indian Communities*), and a study of American Indian self-help housing as part of a larger national study of self-help housing conducted by the Organization for Social and Technological Innovation (OSTI) and funded by the U.S. Department of Housing and Urban Development, plus studies of educational facilities design in England and work-place design in this country. His architectural practice has included the design of high school facilities, youth rehabilitation facilities and housing, plus museum and college campus programming.

# Antonio Guglielmo DiMambro

Antonio Guglielmo DiMambro is an architect, planner and urban designer with experience in two continents. He holds Bachelor's and Master's Degrees from MIT and a Masters in Architecture from I.U.A.V. in Venice, Italy. His expertise encompasses three areas: planning, design, and development of large-scale projects; community participatory process and neighborhood revitalization; and reuse of historic centers and buildings. DiMambro came to the United States first in 1975 as an intern with the Urban Design Group of the New York City Planning Commission. He then returned to Italy, where his work included design and implementation guidelines for a low-income housing development in Rimini; a development program and master plan for the University of Pavia and the design for that University's School of Engineering building; physical and neighborhood anaysis of two city blocks in Urbino, including 15th-century and 16th-century palazzos; and a study for reuse of a 16th-century complex in the city of Calolzio Corte. In 1979 DiMambro again came to this country as urban designer for the Massachusetts Port Authority's Logan Land Use Master Plan. He is currently director of research on tractability in housing and neighborhood form at MIT.

# Thomas P. Bligh, Ph.D.

Thomas P. Bligh completed his studies at Witwatersrand in his native South Africa. He holds an MA in mechanical engineering and a PhD in physics. After teaching for six years at the University of Minnesota in both the Department of Civil and Mineral Engineering and the Department of Mechanical Engineering, he joined the faculty of MIT in 1979 as an Associate Professor of Mechanical Engineering. He is author of numerous articles, papers and reports on energy and underground space and holds four patents on methods and devices for breaking rock and sealing telescoping members.

The unique annual integrated system for heating, cooling and providing electrical power for the new Civil and Mineral Engineering building at the University of Minnesota was designed by Bligh. This co-generation system will combine high-temperature active solar collectors (for heating and for powering an electrical generator) with an air conditioning system that stores winter cold for cooling use in the summer.

Bligh has received several awards for an extra-curricular accomplishment: underwater photography.

# Interview with Edward Allen

Frenette: What were your expectations before you reviewed the projects?

Allen: I was afraid that we would see a lot of buried buildings that were energy efficient, but miserable places to be. Although we did get some like that, I was refreshed when I left after looking at the entries all day because of the high quality that we did see in many of the entries.

It has been my experience that the forces that bear on the design of energy-efficient construction are forces which tend to make a building pretty uninterestng. If the designer allows the technology to take command, the result can be unlivable. One has to be careful not to let the earth-shelter feature run away with the architecture.

Frenette: What particular strengths did you see in the 150-plus projects?

Allen: The most interesting thing I saw was simply a number of good designers who are working very creatively with the earth-shelter concept and making good architecture out of it.

Frenette: Did you get the feeling that the good projects were "good" because they were done by good designers, not because they were earth-sheltered?

Allen: Yes, the best projects were developed by people who could design in any context. In fact, some of the worst projects were those that too narrowly defined earth-sheltering and got lost with the concept.

There were a small number of entries which appeared to be done by people who had not designed buildings before. They got intrigued with the concept of earth-sheltering, but it was plain that, although they understood earth-sheltering well, they didn't know much about architecture. It did appear that the good buildings were by highly trained people.

Frenette: What do you think was the greatest weakness of the entries?

Allen: The tendency of being buried by the concept of earth-sheltering and not being able to find the opportunities in each situation to make totally good buildings. Most entries simply didn't make enough of their opportunities. They were too tied down by the obligation to be earth-sheltered first anc architectural second.

Frenette: What do you think about the idea of having a competition concerning the problem areas, specifically, for multi-family residential?

Allen: It's worth a try. Perhaps you could increase the prizes in that category. However, I don't think you're going to have much success in the multi-family field until clients start asking for more and better earth-sheltered housing.

Frenette: How do you think the students fared compared to the professionals?

Allen: That varies from category to category. In the single-family residential category, we were rather disappointed in the student entries and more pleased and excited by the professional entries. There were several very fine houses by professionals. I don't recall seeing any in the student category that were of the same degree of quality. However, in the multi-family residential category there was the one student scheme that was comparable to anything done by the professionals. In the commercial category, we gave few prizes to students — the chapel and the project for the University of Minnesota.

Frenette: Do you think that it is true that professionals are breaking the new ground in all areas of design and that students in architecture tend to follow more than they lead?

Allen: Yes, I think that's always been true. Students are not encumbered by the kind of "real world" consideration that professionals are. However, they don't have the breadth and length of experience to know all the possibilities.

The professionals are better able to address a particular functional problem that has been troubling people for a long time, or to sense when a problem is not important. They understand that they shouldn't worry about some problems too much because the trouble they cause is not terribly important. I believe in many cases a student will see more limits to design than actually exist. The professional is more likely to have a balanced view of the real and ideal world.

Frenette: What do you think of the bizarre projects that were entered? Strangely, they came more from the professionals rather than from students.

Allen: A trailer house design and one of the single-family projects for the northwest were rather bizarre. I'm glad we didn't give the house a higher prize because there were some real problems in the structural systems.

Also, the design with all the arches was essentially an exercise in energy-overkill. At first it seemed rather beautifully worked out. We got very excited by it at first because it was fairly well drawn. However, the more we looked at it the more it was evident that it was very expensive and that the author of that solution didn't do his homework on the size of the thermal mass that would be optimum for the structure. In addition, he threw together a few too many technical cliches. We started out by being very excited by the logical conclusion that vaults are a very good approach to underground construction. We ended being disappointed because the designer went overboard with technology.

Frenette: My greatest criticism of the projects is that some very modest buildings, like a small office building or a small residence, were intimidated by huge concrete structural systems which were totally out of proportion with the space enclosed or with the amount of energy that could possibly be saved.

Allen: It was evident to me that all the work submitted in the professional category wasn't done by design professionals. It may have been done by contractors.

Frenette: Do you have any recommendations for future competitions?

Allen: You should keep having them. I was skeptical before I saw the entries and wasn't convinced the competition would be worthwhile. After seeing the entries, however, I was impressed with both the volume of work and by the quality that a large percentage of the work had. My suggestion would be to use categories that are attractive and important. Specifically — multi-family residential and research. Single-family residential is always going to be a place where a certain number of people will have a good deal of success.

Frenette: I have a difficult question for you. My only criticism of the jury is that you seemed to bend over backwards to carefully consider rather humble looking structures and often dismissed easily more stylish looking designs, almost without regard to the actual contents of the projects. Will you comment on this observation?

Allen: We were suspicious of schemes that were all "style," and whether we over-reacted or not to that, we were always looking for projects that were well drawn. Good drawings first produced a positive reaction. However, after that positive reaction we became suspicious. We wanted to be sure that the designer really knew what he was doing.

One project that intrigued us was the one that had a central corridor with rooms gong off like flowers on a stem. We certainly admired it very much in concept.

Frenette: It wasn't done totally for energy conservation. The client wanted a house on a point of land that didn't need a house. Earth-sheltering was used, in this case, to conserve the natural quality of the site.

Allen: That's a perfectly valid reason for doing an earth-sheltered building but we, perhaps wrongly, set the criteria at the beginning of the day. First, a project must be good architecture, whether it was buried or not; second, an earth-sheltered project had to make sense for both technical and thermal reasons. Perhaps we should have been more broad minded. We could have looked at designs such as that house and determined that the use of earth-sheltering was for preserving a natural landscape on a beautiful site. On that basis we may have awarded a prize. However, we looked at it from a thermal and cost standpoint and determined that it didn't make much sense. We looked at it for a long time. We argued a lot about it and enjoyed the design a great deal.

Frenette: You might be interested in knowing that we have tried to do very little editing of the personal style of each designer. We have found in our effort to do only light editing that there is almost a direct correlation between the quality of graphics and narrative and the architectural result. Seldom has a poor narrative been part of a terrific project. Often a narrative with gimmicks accompanied a project with energy or architectural gimmicks.

Let me say that you have a fantastic memory and you did a very good job with the jury. Have you discussed the competition since we last met?

Allen: Yes, we have. All of us have better raw material for our own design work after seeing what other people were able to achieve.

We were disappointed that there wasn't more good work in certain categories. We were a bit appalled by some of the amateur work that came in. However, there was enough good work to make us feel it was worth a hard day's work to dig it out and discover it, admire it, and learn from it. All of us left with the feeling that this competition was well worth the effort. It should advance the state-of-the-art.

# Single-Family Residential

"The Single-Family Residential category was, by far, the largest category and I think it continues to serve, and probably always will, as the primary testing ground for new architectural ideas. It is the place where people are willing to take risks. By people, I mean both architects and clients. This is because there is a relatively small amount of money at risk. This is where we saw the largest number of entries and it is here that we saw the most variety, the best and the worst."

Edward Allen

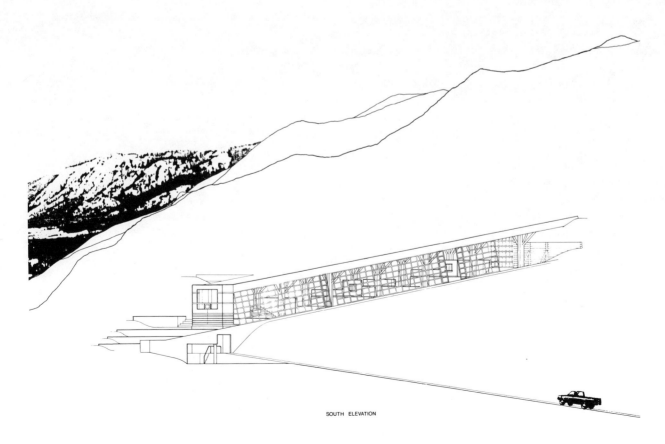

SOUTH ELEVATION

# First Award

## A Farm Home in Central Washington State
### by Robert Hull, Miller/Hull Partnership, Seattle, Washington

The project was to design a house for a family of four on a small farm site. The client expressed interest in a house that would be as self-sufficient in energy use as possible including passive solar design. Because of the family's living and entertainment style it was requested that the design focus on a greenhouse that could function as living areas and solar collection space.

The site is located in Central Washington State in a harsh climate that is hot and dry in the summer and cold in the winter. Temperatures vary from 110°F to extended periods of freezing. Irrigated farming is the main industry in the area.

It was decided to "go underground" early in the design process. The winter heat loss and summer heat gain calculations were conclusive. In addition, the resultant structural system of concrete walls and slabs provides mass to the house that can be utilized to store the solar gain from the greenhouse and to minimize temperature fluctuations. The constant earth temperature of 45-50°F, along with buried culverts/ducts, will be used for summer cooling.

Functions have been skewed in plan to create a centrally focused greenhouse surrounded by the house proper. The concrete floor slabs tier up from 12 feet at the south to 8 feet at the bedroom/kitchen. Entry to the house is by tunnel.

The final solution was generated because of the inherent restrictions of a greenhouse (they're fine for winter but overheat in the summer). The architects felt that basically it would be best to "take away" the greenhouse in the summer by using commonly available wood framed residential garage doors as the variable skin. The doors function as a greenhouse wall in the winter and then roll up completely in the summer, opening the greenhouse to the outside. They then become the summer shading device for the

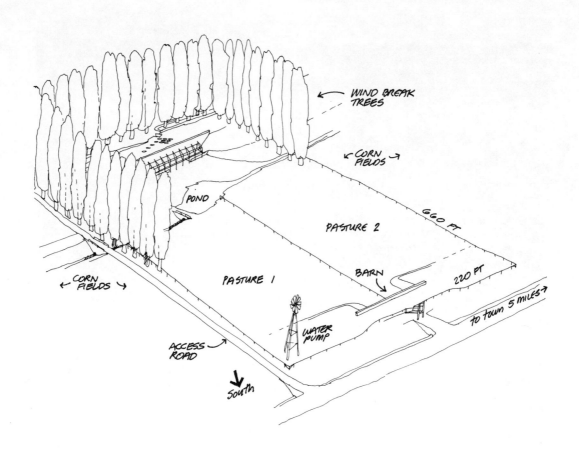

stationary glass roof without destroying the vertical angle of view from the house. In the summer, the interior skin of the house at the bedrooms, kitchen and family room functions as the lockable and insect proof layer and the greenhouse becomes an outside porch. In the winter it becomes part of the enclosed interior space.

This house is an attempt to incorporate energy concerns into the design process and emphasizes the architect's responsibilities to a unified design approach without resorting to the cliches of solar techniques.

Floors are concrete slab, etched and polished. Exterior walls are cast in place concrete and interior walls are gypsum board. The roof is cast-in-place flat concrete slab with inverted beams. Lightwells are precast concrete pipe.

Mechanical systems include passive solar with wood stove backup heating and radiant task heating.

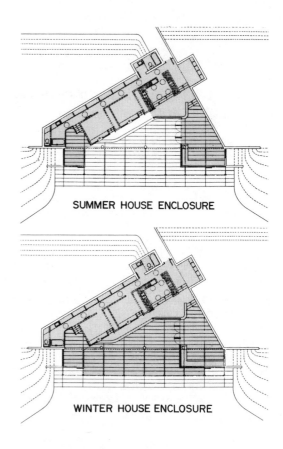

SUMMER HOUSE ENCLOSURE

WINTER HOUSE ENCLOSURE

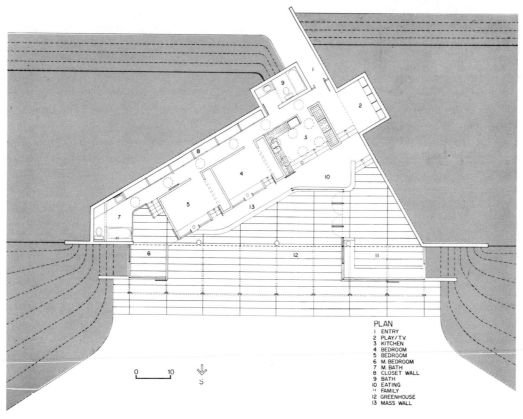

0    10

S

PLAN
1  ENTRY
2  PLAY/ T.V.
3  KITCHEN
4  BEDROOM
5  BEDROOM
6  M. BEDROOM
7  M. BATH
8  CLOSET WALL
9  BATH
10 EATING
11 FAMILY
12 GREENHOUSE
13 MASS WALL

18

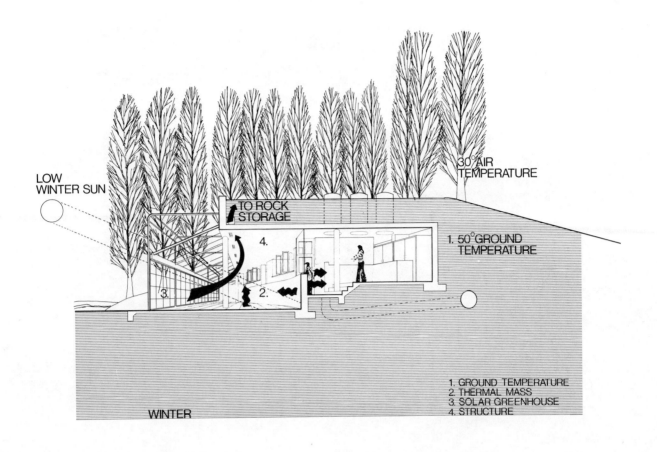

LOW
WINTER SUN

30° AIR
TEMPERATURE

TO ROCK
STORAGE

1. 50° GROUND
TEMPERATURE

4.

3.

2.

1. GROUND TEMPERATURE
2. THERMAL MASS
3. SOLAR GREENHOUSE
4. STRUCTURE

WINTER

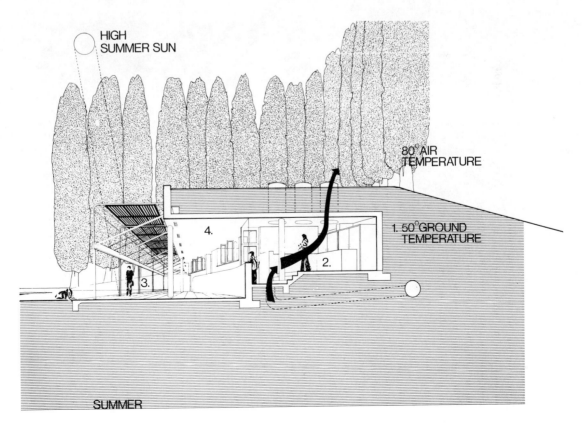

HIGH
SUMMER SUN

80° AIR
TEMPERATURE

1. 50° GROUND
TEMPERATURE

4.

3.

2.

SUMMER

ISOMETRIC

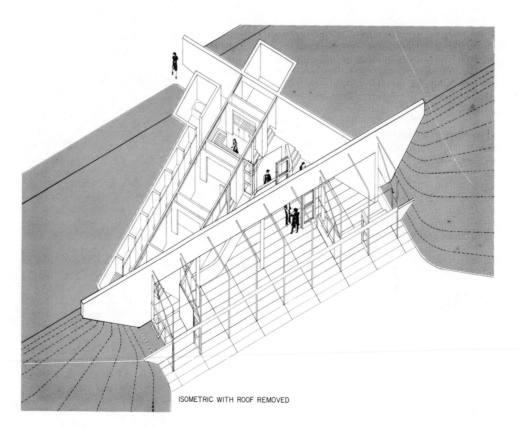

ISOMETRIC WITH ROOF REMOVED

# Merit Award

## A Ranch in the Mountains of Northern Washington State
by Robert Hull, Miller/Hull Partnership, Seattle, Washington

The task was to design a house for a couple on a steep ranch site in the Cascade Mountains of Northern Washington State. Early input by the clients expressed interest in passive solar and earth-sheltered design as well as a desire to have the house recreate the traditional feelings of a Northwest lodge.

Because of its mountainous location the climatic conditions are extreme. Winter snows of three to four feet are common, accompanied by extended periods of freezing. Summer temperatures can range into the nineties. The lower portion of the site is forested with evergreens. Consequently, the house will be located high above the treeline in order to gain good solar exposure and a panoramic view into the valley below and the mountain range beyond.

The concept developed out of a difficult site constraint. The steep site falls off to the southwest, yet the solar collector needed to face due south for maximum efficiency. This constraint became the generator of the solution. Skewing the solar collector to the south and consequently up the hill, reduced the difficult hill incline to an angle that is walkable for a ramp. The ramp and collector became the access route for the house. The house became a series of tiers with views out over the valley below. The lower tier is the living/entertaining space and is dominated by a large scaled fireplace and wood furnace.

The desire by the clients for a "lodge" type house required the exploration of wood as a possible structural system. However the heavy roof and retaining wall is crenelated to avoid horizontal sheltered structures necessitated keeping spans as short as possible. The structural system chosen reduces all spans to a maximum of eight feet. In order to avoid a grid of columns at eight foot centers, radiating tree columns were chosen to collect the loads from the roof. The rear

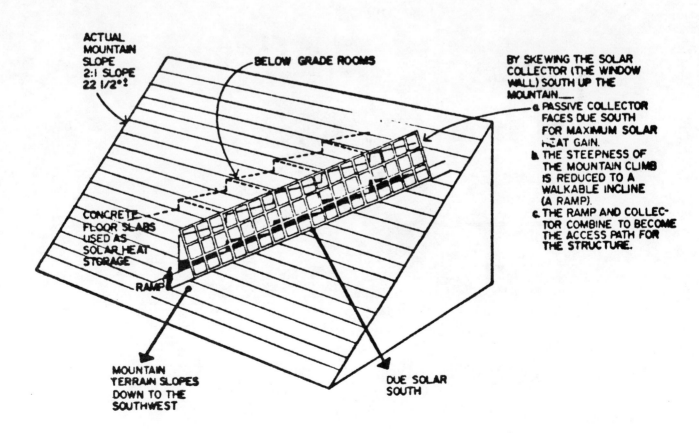

ACTUAL MOUNTAIN SLOPE 2:1 SLOPE 22 1/2°!

BELOW GRADE ROOMS

BY SKEWING THE SOLAR COLLECTOR (THE WINDOW WALL) SOUTH UP THE MOUNTAIN___

a. PASSIVE COLLECTOR FACES DUE SOUTH FOR MAXIMUM SOLAR HEAT GAIN.

b. THE STEEPNESS OF THE MOUNTAIN CLIMB IS REDUCED TO A WALKABLE INCLINE (A RAMP).

c. THE RAMP AND COLLECTOR COMBINE TO BECOME THE ACCESS PATH FOR THE STRUCTURE.

CONCRETE FLOOR SLABS USED AS SOLAR HEAT STORAGE

RAMP

MOUNTAIN TERRAIN SLOPES DOWN TO THE SOUTHWEST

DUE SOLAR SOUTH

retaining wall is crenulated to avoid horizontal spans greater than eight feet. A local log cabin manufacturer will supply the ten inch round milled log components.

Using roll-up wood garage door sections turned vertically to form double glazing, the solar collector functions as a window wall with full-height operating sections for natural ventilation. This system is used throughout the structure as an inexpensive glazing system. The mass of the exposed concrete floor slabs, directly behind the glazing, functions to store the solar thermal heat.

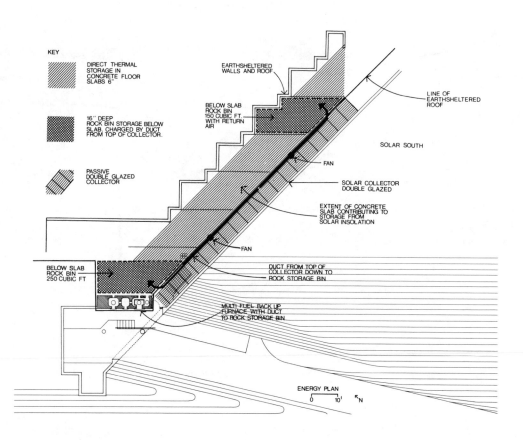

KEY

DIRECT THERMAL STORAGE IN CONCRETE FLOOR SLABS 6"

16" DEEP ROCK BIN STORAGE BELOW SLAB, CHARGED BY DUCT FROM TOP OF COLLECTOR.

PASSIVE DOUBLE GLAZED COLLECTOR

EARTHSHELTERED WALLS AND ROOF

LINE OF EARTHSHELTERED ROOF

BELOW SLAB ROCK BIN 150 CUBIC FT. WITH RETURN AIR

SOLAR SOUTH

FAN

SOLAR COLLECTOR DOUBLE GLAZED

EXTENT OF CONCRETE SLAB CONTRIBUTING TO STORAGE FROM SOLAR INSOLATION

FAN

BELOW SLAB ROCK BIN 250 CUBIC FT

DUCT FROM TOP OF COLLECTOR DOWN TO ROCK STORAGE BIN

MULTI-FUEL BACK UP FURNACE WITH DUCT TO ROCK STORAGE BIN

ENERGY PLAN
0        10'        N

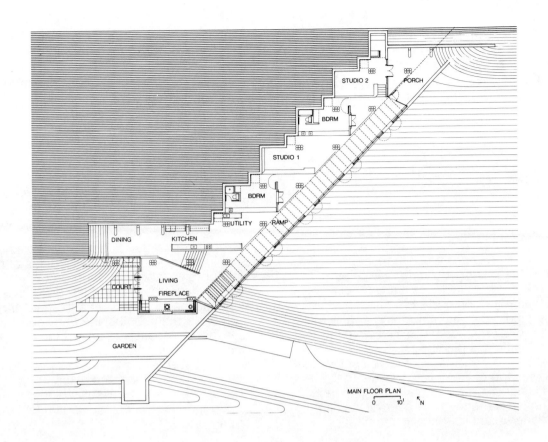

DINING

COURT

GARDEN

KITCHEN

UTILITY

LIVING

FIREPLACE

RAMP

BDRM

STUDIO 1

BDRM

STUDIO 2

PORCH

MAIN FLOOR PLAN

0    10'    N

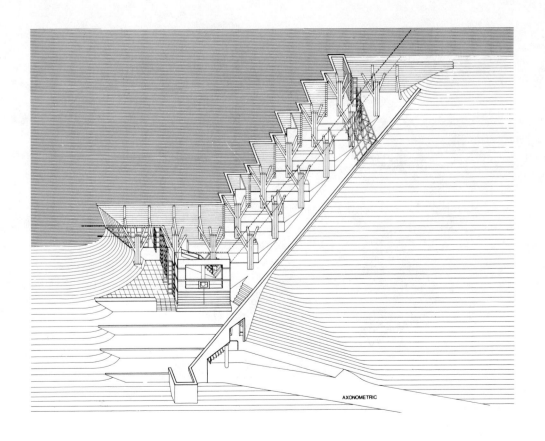

AXONOMETRIC

# Honorable Mention

## Artist Residence for Boulder, Colorado by Geoffrey Harris, Designer, The New and Improved Studio, Boulder, Colorado

The challenge was to design a primary residence for an artist/builder who is sensitive to the impact of a building on the landscape, who desires the building to occupy a position in harmony with the land and who wants the building to be energy efficient. The program includes a space for passive solar collection and storage, two bedrooms, study, spa, kitchen, dining and living room and two full bathrooms, "which would indeed be rooms, not plumbed closets." A further consideration is an outside sculpture display area to take advantage of kinetic response to natural energy forms and environmental changes on the site. The house is a conscious attempt to utilize all the established practical benefits of earth-sheltering and combine these with a sculptural earth form — house as "earth art."

The site is a small lot of about one-third acre, generally flat, set back about 100 feet from a ridge overlooking the city of Boulder. The site has excellent solar exposure and a high view of a rock formation known as the Flatirons.

Since the owner is intrigued by earth forms, ruins and the combination of the two in the process of time, an earth-sheltered house is the design response. The design places the house close to the center of the lot, extending along an east/west axis approximately six feet below existing grade. Earth is bermed against the portions of the walls which project above existing grade and two feet of earth covers the roof of the entire structure. Penetrations through the earth-sheltering occur at both ends, at the baths, the entry court, the dining area, above the kichen, and at the entry to the garage. One approaches the house along a path on the east side and descends into the entry court. The entry doors are glass which allows a view along the east/west axis of the house through the loggia, and out again into the west court.

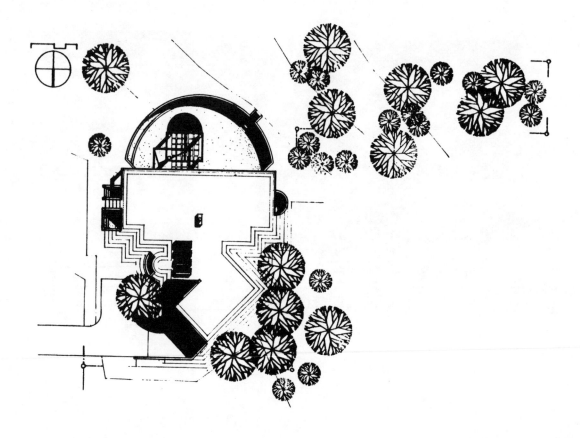

All primary spaces open to the loggia which is the main circulation element of the house as well as the passive solar collection space. The loggia contains 14-18 inch diameter water filled columns eight feet high and arranged in pairs centered on the structural bays. Each column also contains a light source centered within the water. The design alludes to examples of classical architecture, in effect a "solar temple."

The loggia in turn opens to a large semi-circular meditation space/sculpture garden 7'-4" below existing grade. In the space, concrete columns support a wood frame outline of a gable. This pavilion contextually links the house to its neighbors, defines an exterior "room" along the north/south axis of the house, and provides a scaffold to support any combination of awnings, wind chimes, banners or soft sculpture. The south sculpture garden elevation of the house is almost entirely glass, fronted by two rows of regularly spaced concrete columns which support the earth covered overhang and reflect the ordered structural grid within.

Walls are eight inches of reinforced concrete. Columns are 14 inch diameter, cast-in-place concrete. Roof is 14 inch Trus-joists with four inches of concrete slab. The floor is a five inch insulated concrete slab in the loggia with thermal breaks between adjacent four inch slabs in other rooms. Finishes are exposed concrete walls, exposed roof trusses (painted), quarry tile floors in most public areas, carpet in most private areas, rubber flooring in the kitchen. The exterior is of indigenous grasses, gravel or rock chips in courts and the sculpture garden.

Collected heat is withdrawn from the loggia by small electric fans located high in the wall between the loggia and the adjacent rooms. Transom windows above each room's French type doors are operable to allow natural ventilation. The French doors are glazed with reflective glass for additional privacy. The kitchen skylight utilizes two operable roof windows for ventilation, which have roll up insulation.

The structure has four inches of rigid insulation on the roof, two inches on the walls, and one inch below the loggia slab. Exterior frame walls are 2' x 8' with full fiberglass insulation. The roof slab incorporates thermal breaks between all interior and exterior spaces. All glazing is an insulating type.

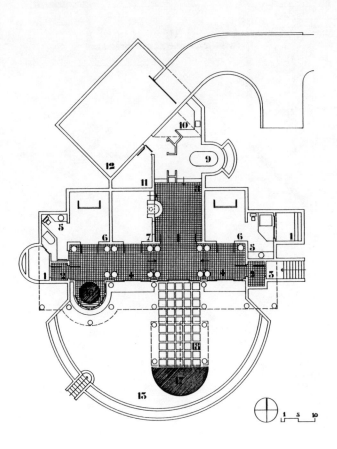

**LEGEND:**
1. Courtyard
2. Airlock
3. Entry
4. Loggia
5. Bathroom
6. Bedroom
7. Study
8. Living Room
9. Dining Room
10. Kitchen
11. Utility/Mechanical
12. Garage/Shop
13. Sculpture Garden
14. Solar Hot Water Collectors
15. Ventilating Skylight
16. Hot Tub
17. Reflecting Pool
18. Pavilion

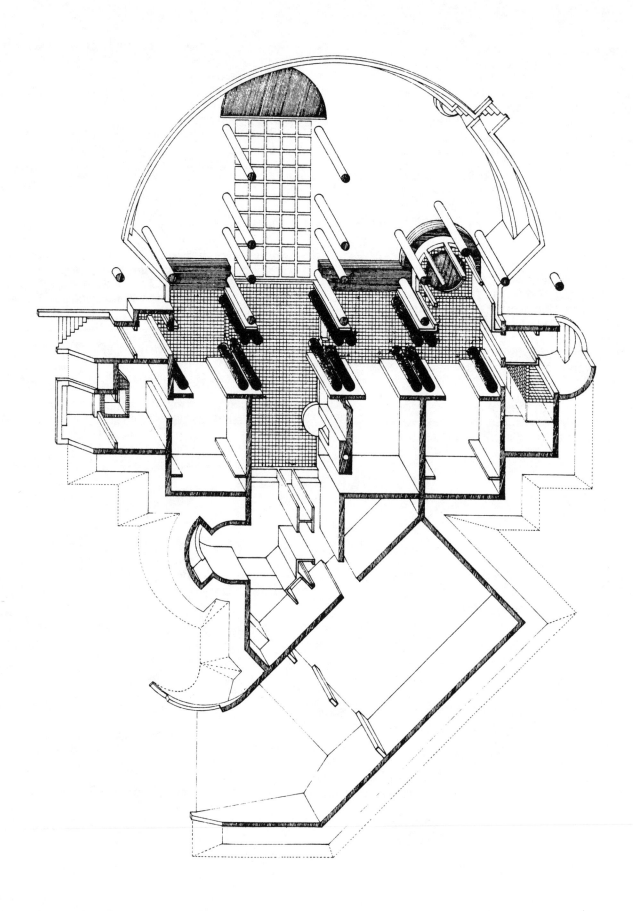

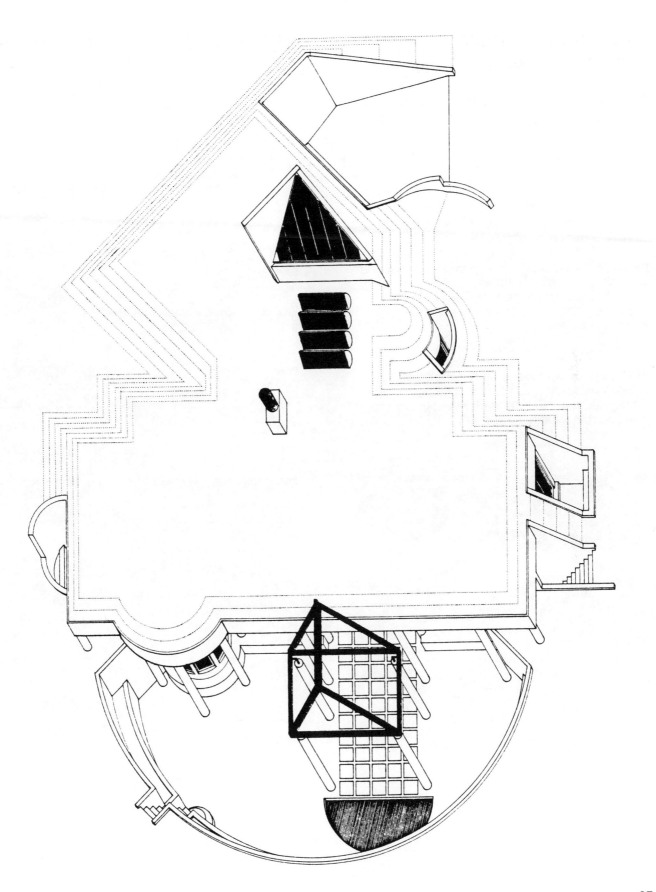

# Honorable Mention

## A Classical Residence for Manhattan, Kansas
### by Gordon Ashworth, Architect, Kansas State University, Manhattan, Kansas

This is a proposal for an earth-integrated single-family residence on a southwest sloping site in the Flint Hills region of Kansas. It is designed with the belief that it is possible to build a house that can combine many of the advantages of energy-efficient design with the usual functional and psychological needs of a middle-income suburban family.

Many energy-efficient buildings have a self-conscious pioneering quality about them which this designer believes is un-nerving to many people. Underground houses in particular often do not seem to offer the qualities of living which most people expect.

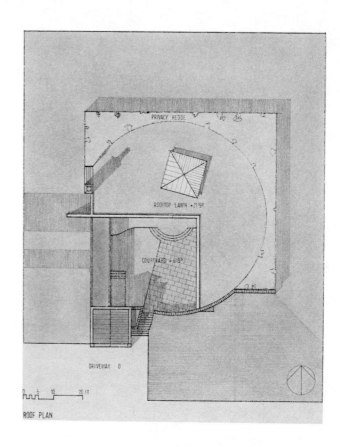

ROOF PLAN

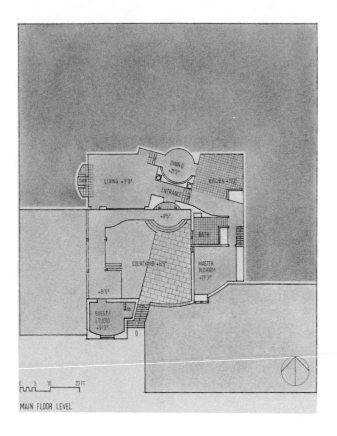

MAIN FLOOR LEVEL

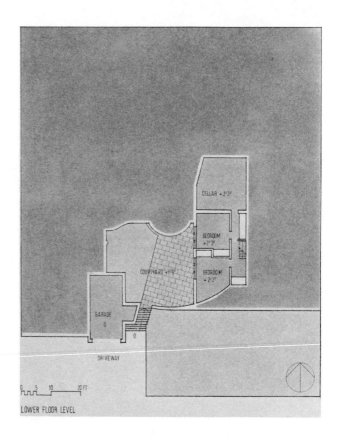

LOWER FLOOR LEVEL

This design attempts to reconcile these opposites by using a traditional L-shaped courtyard house and adapting it to an earth-integrated situation. The courtyard house has been a familiar prototype for centuries and has been used in northern and southern climates with great success. In cold weather, it can provide shelter from winter winds and in hot weather it can provide shade. Other general advantages are that the L-shape can enable the house to be easily zoned into communal and private realms within the house. The courtyard forms a private outdoor space with minimum fences. The L-shape can be used in linked multi-family situations with zero-lot zoning and is easy to adapt to most topographical situations. The courtyard house is particularly adaptable to earth-integration because of its single-aspect characteristics, with circulation at the rear and prime space at the front.

The house is 1,900 square feet (excluding garage and cellar) and is three bedrooms with a studio/ guest room included. The studio is free-standing from the rest of the house (connected by a covered way) and is set above the garage, with a pitched roof, to give the familiar impression of a traditional two-story house from the approach side. A flight of steps leads from the driveway up to the courtyard, which is set approximately half a level above the garage, to give privacy and views. An axis from the garage and steps leads to the main formal entrance into the house and terminates in a circular dining space, which is the most formal part of the house, associated with the family ritual of dining. This space is set up a few steps to heighten these formal aspects and is enclosed in an aedicula form by columns and a pyramidal roof, which becomes the vestige of the traditional roof, seen from outside. Glass blocks are used to let light into this space through a clerestory light. Light also infiltrates the kitchen and rear of the living room from this source.

The other areas of the house are traditional in character, with a living space to one side and bedrooms and bathroom in the eastern wing. All these rooms look into the court and the master bedroom also has a bay window which looks out into the landscape.

The materials are poured-in-place concrete for the work underground and concrete blocks elsewhere. Facing is stone veneer on the garage/ studio and white stucco in the courtyard. The front door and windows are surrounded by green ceramic tiles.

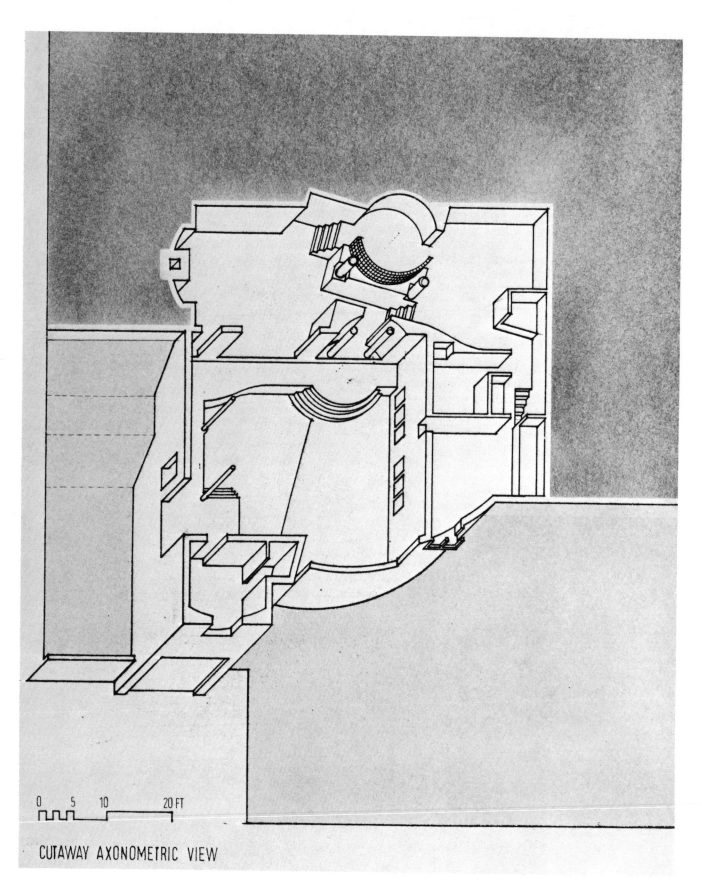

0    5    10              20 FT

CUTAWAY AXONOMETRIC VIEW

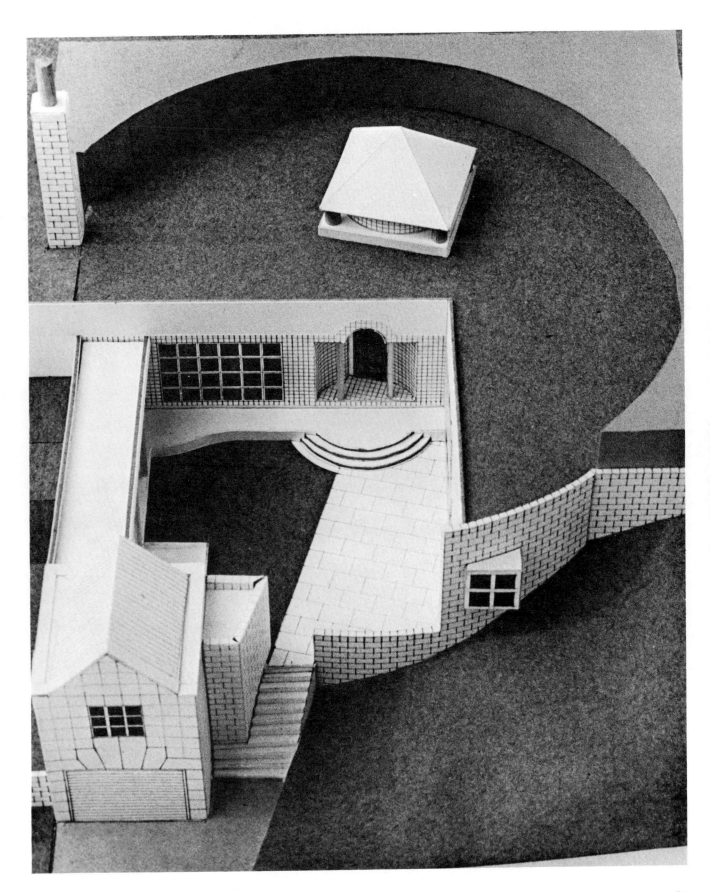

# Honorable Mention

## House on a Pond near Ponca City, Oklahoma
by David M. Gilmore, Student, Oklahoma State University, Stillwater, Oklahoma

The site is the northeast bank of a four acre pond in Ponca City, Oklahoma. The water level of the pond is controlled, so it does not rise more than six inches above the zero level shown on the site plan. The pond is approximately 15 feet deep at low point. It is stocked with fish and will be used for recreational activities. The 1,250 square foot Marshall Island, to the southwest of the site, is wooded and will be used for camping, picnicing, etc. To the west of the site is a densely wooded area to be left as it is. Access to the site is from a public two-lane road to the northeast.

The orientation of the house is to the south and westward views of the pond, an island, and a densely wooded area. The trees will be used for their shading characteristics.

When approaching the house two structures are visible — the garage, which is almost obscured by trees, and the entry, which is more clearly visible. The lower level or living area is 2,700 square feet, organized around the kitchen with the main views to the pond and Marshall Island.

During months when cooling is necessary, shading devices are used along with the natural shading of trees. Manually-operated sliding panels are used on the west side of the living area along with another sliding shade at the terminus of the living space. Natural ventilation is achieved by operable windows, allowing cool air from the water into the house. Hot-air-powered ventilating fans draw air up through the exposed cuts in the entry way. Glass block walls are used to allow light into the bedrooms while still achieving privacy. The earth covering on the roof and earth on the north side of the house add additional insulation.

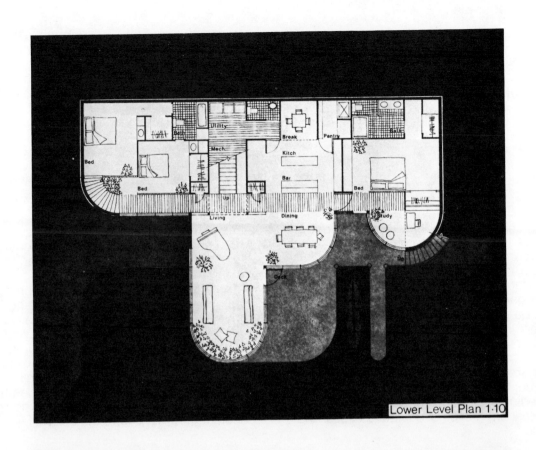

Lower Level Plan 1·10

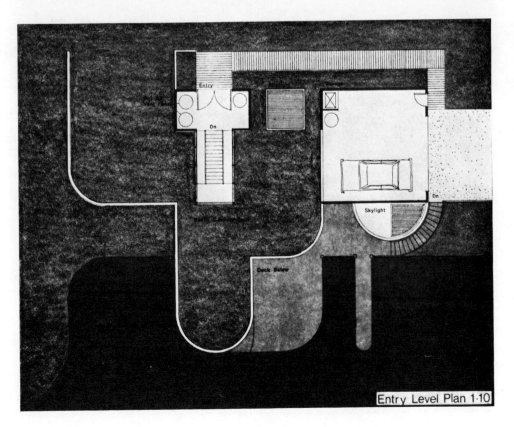

Entry Level Plan 1·10

33

# Honorable Mention

## House on a Pond near Ponca City, Oklahoma
### by Steve Wherley, Student, Oklahoma State University, Stillwater, Oklahoma

The intent of the project was to design an earth-sheltered single-family residence for a site in Ponca City, Oklahoma. The client requested that the house contain a family/living room, formal dining room, kitchen, eating nook, study, three bedrooms, three baths, double garage and storage for one boat. The client also asked that the house be organized around the kitchen to fit their present life style.

The site is located at 36° north latitude. The south side of the site is bordered by a four acre pond. The pond contains a small island, on which there are two medium sized deciduous trees. To the east and west are groves of deciduous trees. The

access road runs along the northeast edge of the site which is 60 feet wide in the north-south direction and 79 feet deep from the pond to the property line in a northeasterly direction. There is a six foot building setback line on the north border of the site and a 25 foot setback on the northeast edge of the site. The site slopes from a grade of ten feet above the water along the road to the water's edge making it an ideal location for an earth-sheltered house. The best view from the site is slightly west of south across the pond.

Due to the site restrictions and its adaptability to accommodate an earth-sheltered house, a linear scheme was chosen. All rooms open to the south, taking advantage of the views, with circulation along the north wall. The entry is a procession of openings in white stucco walls leading from the driveway to the greenhouse, along a wooden deck. Once inside the greenhouse, which also acts as the formal entry, one is on a bridge looking into an interior space that visually links the indoors and the outdoors. A sculptural spiral stair connects the bridge to the living areas. This bridge continues through the greenhouse to the deck at the west end of the house.

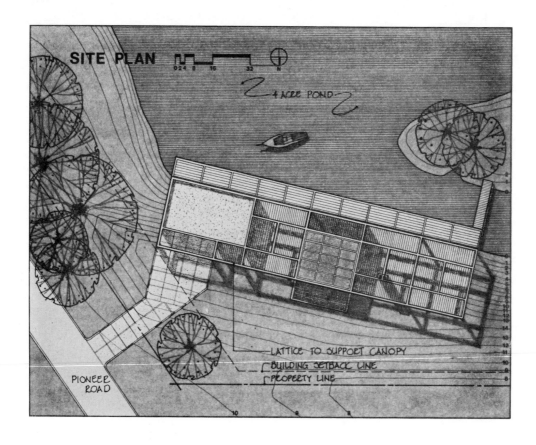

A louvered sun screen is used on the south wall to shade the house from the sun. The sun angle is 78° during summer solstice and 30° during winter solstice. The change between the heating and cooling season occurs at a sun angle of 54°. The louvers block out the sun at angles of 54° and greater during the summer cooling months, the critical period of the year. The sun screen which allows visibility of the sky, allows sun penetration at solar angles of less than 54°. In addition, the deciduous trees on the island are also employed as shading devices for the living room and master bedroom.

The greenhouse/solarium is used to aid natural ventilation and operable windows in the south wall can be opened in the evenings to allow the natural breeze from the pond to enter the house. Air exits through the well to the north of the greenhouse and the ceilings in the house are sloped to aid natural ventilation and to permit deeper sun penetration. The east and west walls of the greenhouse are made of white stucco to reflect light into the dining and living rooms. The floor of the greenhouse is made of heat-absorbing clay tiles, which radiate heat into the living areas during the heating months.

The lattice above the ground level will be used to support a canopy during the summer months. This canopy will shade the ground above the roof during the cooling months and allow air to move beneath the canopy to further cool the ground insulating the roof of the house. During the winter heating season, the canopy would be removed to allow the sun to heat the ground above the roof.

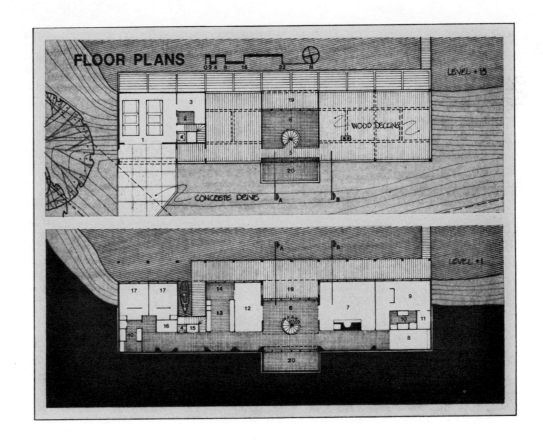

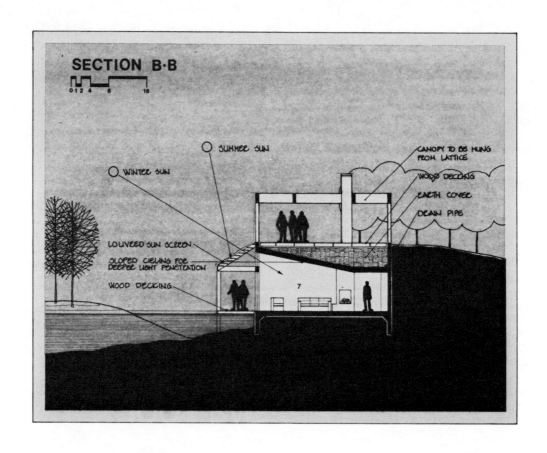

**SECTION B·B**

0 1 2 4    8        16

WINTER SUN

SUMMER SUN

CANOPY TO BE HUNG FROM LATTICE

WOOD DECKING

EARTH COVER

DRAIN PIPE

LOUVERED SUN SCREEN

SLOPED CIELING FOR DEEPER LIGHT PENETRATION

WOOD DECKING

7

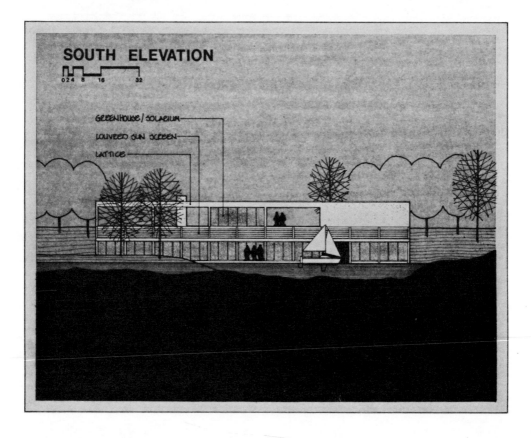

**SOUTH ELEVATION**

0 2 4 8    16       32

GREENHOUSE / SOLARIUM

LOUVERED SUN SCREEN

LATTICE

# Honorable Mention

## Low Cost Residence for a South Facing Hillside
by Doug Graybeal, Graduate Student, Aspen, Colorado

This small earth-sheltered, passive solar house is designed to solve the typical needs of a low-income family. It is planned for any south facing hillside and to take advantage of the thermal mass and temperature lag of the surrounding earth for energy conservation.

At only 1250 square feet, it is a very simple yet functional plan with an open kitchen, dining, and living area. This main living area with its large south glazing creates an ideal direct gain solar space and helps create a spacious feeling by opening the house to the exterior. The outdoor patio off this space also helps to visually expand the interior to the exterior. Further natural lighting for the north rooms is provided by the upper clerestory.

The structure of the house is poured-in-place concrete with precast concrete planks forming the lower roof. The upper roof, because of its slope, is wood post and beam construction.

To help save energy, the following points have been incorporated into the design: the concrete mass is fully insulated and separated from the exterior with thermal breaks; all plumbing and electrical outlets are located away from exterior walls; the garage is a thermal buffer space; the roof, north, east, west and part of the south wall are insulated and protected by earth berms; all plumbing and mechanical fixtures are centrally located; heating is provided by direct solar gain and a heat-recovery forced-air system; cooling is by natural ventilation and natural thermal lag.

**WEST ELEVATION**

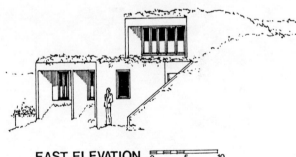

**EAST ELEVATION** 0  5  10

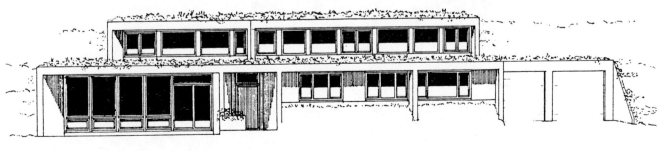

**SOUTH ELEVATION** 0  5  10

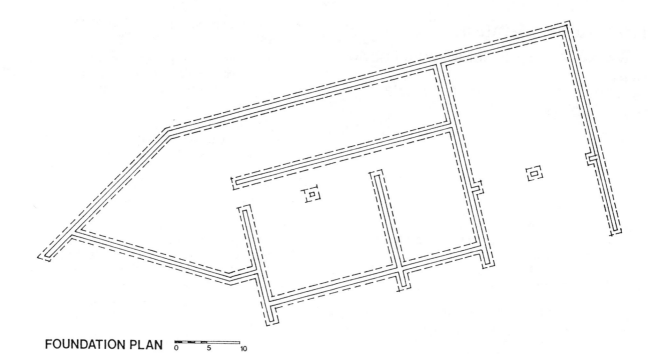

**FOUNDATION PLAN** 0  5  10

NORTH

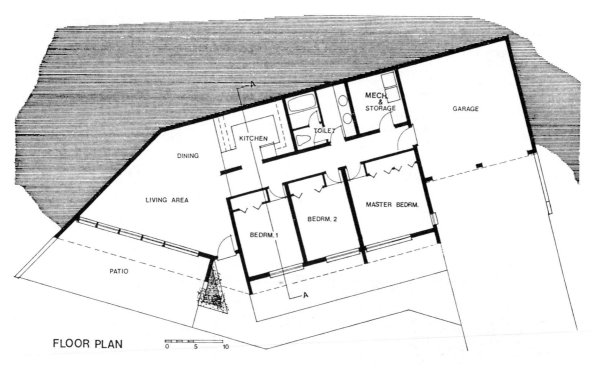

**FLOOR PLAN** 0  5  10

NORTH

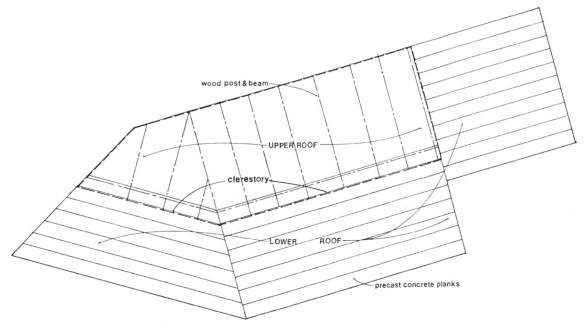

## ROOF STRUCTURAL PLAN

0    5    10

wood post & beam

UPPER ROOF

clerestory

LOWER ROOF

precast concrete planks

NORTH

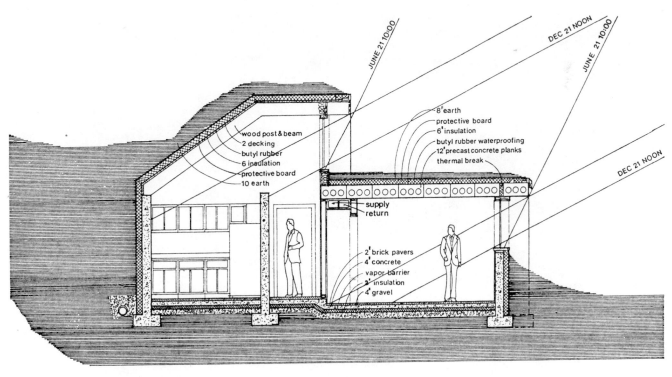

JUNE 21 10:00

DEC 21 NOON

JUNE 21 10:00

DEC 21 NOON

8" earth
protective board
6' insulation
butyl rubber waterproofing
12' precast concrete planks
thermal break

wood post & beam
2 decking
butyl rubber
6 insulation
protective board
10 earth

supply
return

2' brick pavers
4' concrete
vapor barrier
2' insulation
4' gravel

## SECTION A

0    5    10

## Multilevel Residence for Kalamazoo, Michigan
by Gunnar Birkerts, President, Gunnar Birkerts & Associates, Birmingham, Michigan

This dwelling unit is designed for a doctor and his writer-poet wife, requiring a living area, two studio spaces, a master bedroom, kitchen, dining room and an exercise space.

The site has been carefully selected to allow the construction of an energy-conscious dwelling. The steep slope, facing south, permits an almost ideal approach to passive solar design.

The central open stair connects a two-story wing on the west, and a three-story wing to the east. The house is entered on the top level. All the living spaces are fully opened to the south and fully earth protected towards the north. The greenhouse zone at the ground level extends vertically across the south exposure. The concrete, externally-insulated walls are internally warmed by circulating warm air from the greenhouse. In the summer the greenhouse has outside ventilation. At the skylight level, water pipes are embedded in the concrete mass to heat and supply domestic water all year round. Electric radiant heat coils provide tempering of the greenhouse zone during the cold winter months. Concrete walls, not earth covered, are insulated and clad in wood siding on the exterior. In summer, shading is provided by the deciduous trees on the south. Heat loss during the night cycle is minimized by the insulated drapes on the south glass exposure. Landscaping, except the shade trees, will be developed by the owner after occupancy.

The owner is fully aware of and sympathetic to the resulting consequences — three level living, maintenance of glass and plant material and operation of shading devices and skylight shuttering.

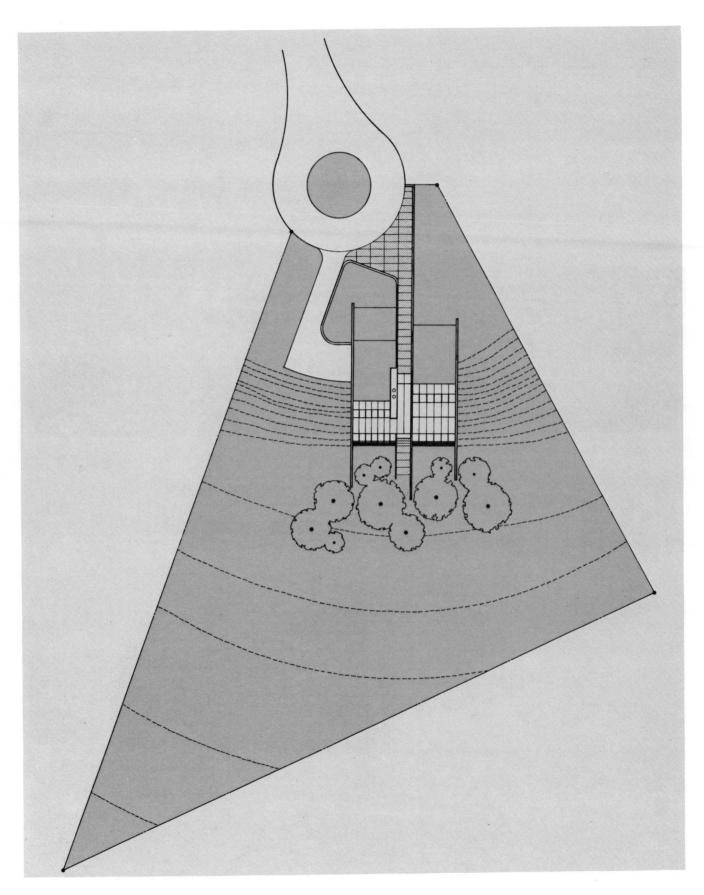

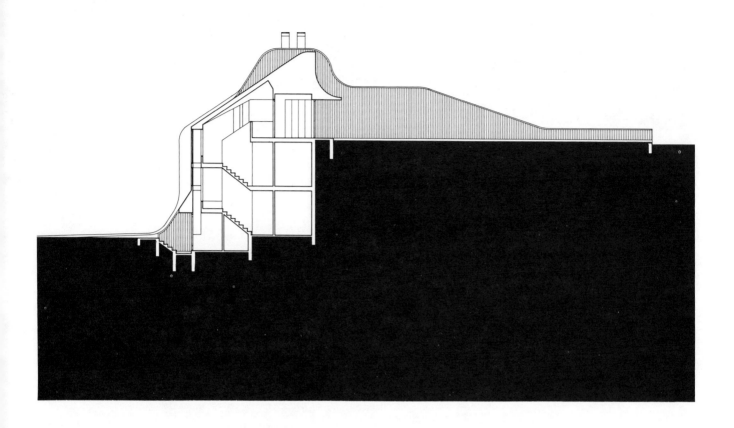

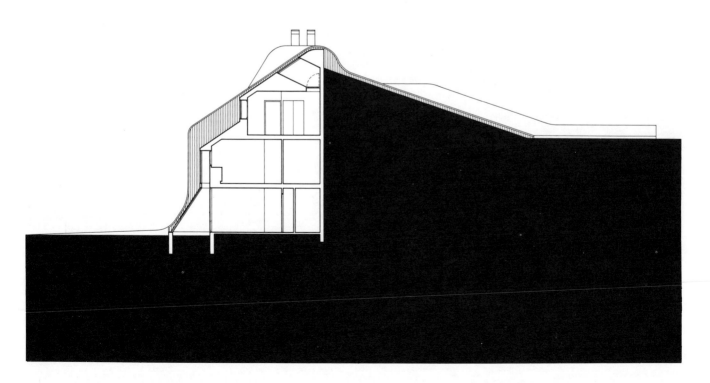

42

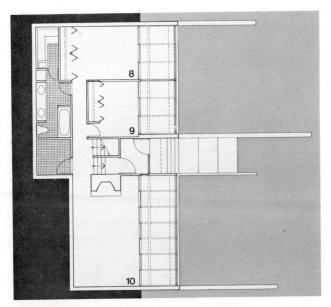

## Level 1 Plan

0   5               25

8   Master Bedroom
9   Studio
10  Exercise Room

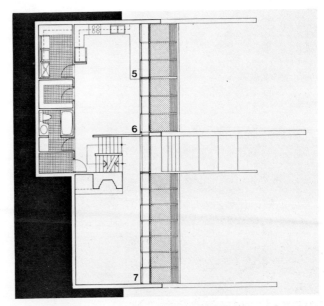

## Level 2 Plan

0   5               25

5   Kitchen
6   Dining Room
7   Living Room

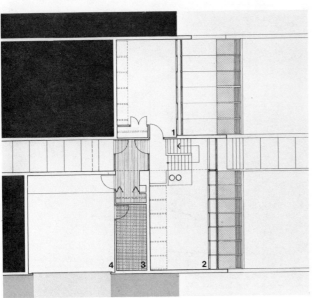

## Level 3 Plan

0   5               25

1   Studio
2   Open to Below
3   Storage Room
4   Garage

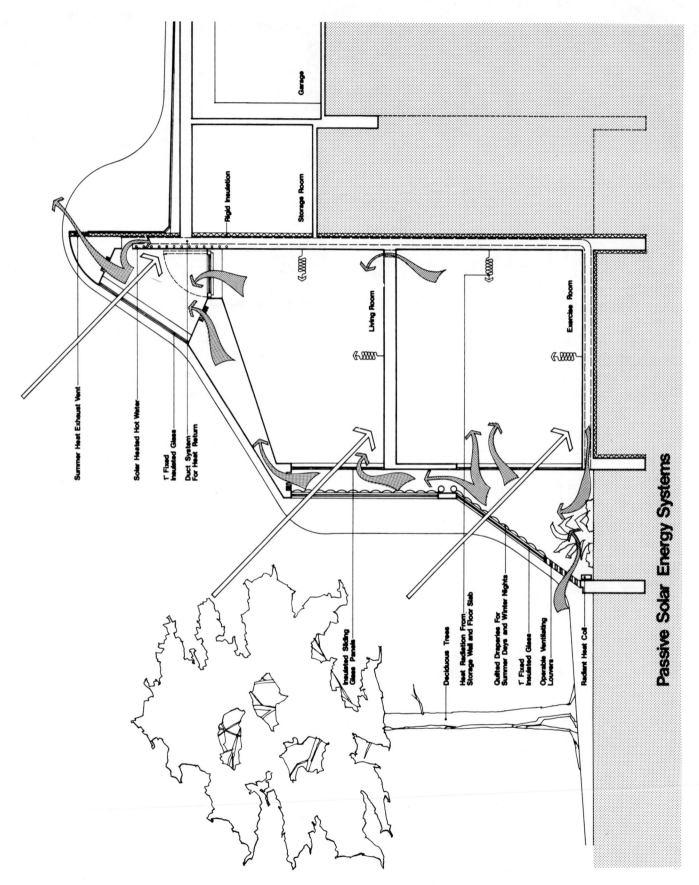

**Passive Solar Energy Systems**

Garage

Rigid Insulation

Storage Room

Living Room

Exercise Room

Summer Heat Exhaust Vent

Solar Heated Hot Water

1" Fixed Insulated Glass

Duct System For Heat Return

Insulated Sliding Glass Panels

Deciduous Trees

Heat Radiation From Storage Wall and Floor Slab

Quilted Draperies For Summer Days and Winter Nights

1" Fixed Insulated Glass

Operable Ventilating Louvers

Radiant Heat Coil

## A Multi-Level Residence Formed by Efficient Structural Shapes
by Alfredo DeVido, Architect, New York, New York

The house is unusual in its use of the curved forms, expressive of the cave-like character of underground houses. These forms have special structural integrity and serve to lighten the otherwise heavy walls that are necessary in underground construction.

This single-family house utilizes a corbelled and curved brick structure. Three tiers of south-facing windows, about 300 square feet, bring light and solar heat into the house. The first tier shelters a greenhouse and an air-lock entry. The second tier rims the south side of the master bedroom, and the third tier brings light into the inner recesses of the house.

Above the house is a hot-air solar collector, which provides domestic hot water and additional space heat. It is ducted to a rock storage bed beneath the floor of the house. Energy is conserved via skylids and night insulation.

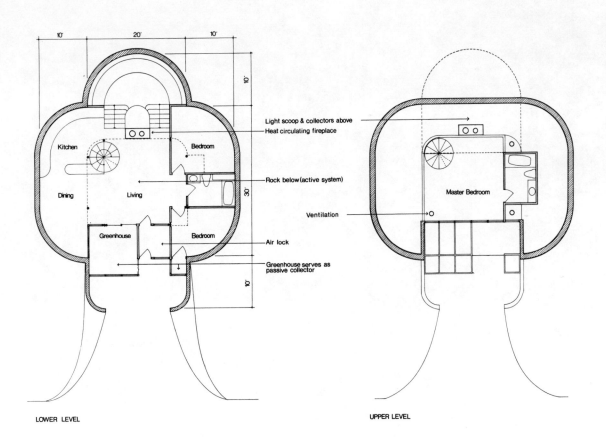

LOWER LEVEL

UPPER LEVEL

Bill Phillips

## Residence Overlooking a Valley Near Portland, Oregon
by Tedd Chilless, President, Chilless Nielsen Architects, Portland, Oregon

The design of this single family residence was predicated on the difficult grades, stringent setback and height requirements imposed on a choice south facing lot which was vacant for many years. This lot provided an ideal vehicle for an energy-smart home of tomorrow. Overlooking a valley, it is located in a rolling, well established residential area with a wide variety of housing styles. Energy efficiency was top priority for the structure. Both active and passive measures were used.

In this 2,162 square foot earth-sheltered home, 52% of the total heating demand will be supplied by passive solar elements. To achieve this result,

the following methods were employed. Full height earth berms are used at the north wall and partial berms at the east and west wall. Two feet of earth covers the roof area. South facing glass, a greenhouse as well as skylights are used for light and heat transmission to the floor. Thermal mass walls and floors with perimeter insulation are used throughout.

The south-facing greenhouse provides a weathertight dual-door entry that traps heated air. Use of operable glazed openings and a fan allows air to be vented outside or recirculated to various interior spaces. The raised living space takes advantage of the view through the greenhouse foliage. Glass block becomes a repeated theme. It is used to connect, but differentiate between the skylit kitchen and the living area, and to enclose the hot tub space off the master bedroom.

Domestic water will be heated by three different methods. The primary source of heat to a central storage tank is solar panels. A water jacket in the wood burning stove will circulate and reheat the water when it is in use. In addition, auxiliary heating will be provided by a gas-fired water heater.

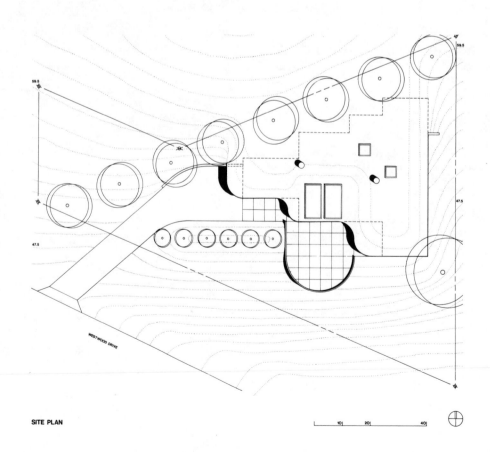

SITE PLAN

A gas fired boiler and fin tube radiation system is a backup to the wood stove and is expected to supply a majority of the 18.32 million BTUs annual heating load. The water in the system will return through the central storage tank to boost the water temperature before returning to the boiler.

The exterior walls are constructed of reinforced concrete with rigid insulation, and wood studs with both rigid and batt insulation. Drywall is the finish on the interior wood stud partitions. Quarry tile finishes the floors in the solar collection areas, and extends out to the terrace. Carpet is used elsewhere. The roof consists of laminated vertical wood decking with sheathing, building paper and a topping slab. White stucco is the exterior finish.

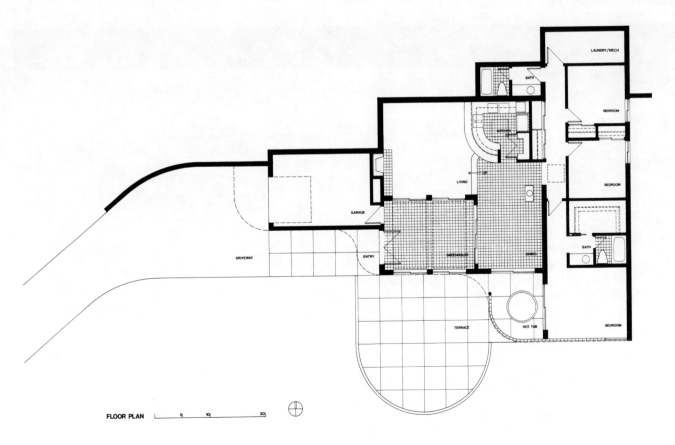

FLOOR PLAN

# House in a Circular Berm
by Robert A. Banbury, Student, Ohio
State University, Wester, Ohio

The house occupies a segment of a landscaped berm atop a hill in Central Ohio. It is situated at the edge of a dense forest with vistas of other hilltops framed by the trees.

The solar panels (for heating the pool and house) are placed upon the berm, away from the house, to facilitate maintenance and repair, and to free the bermed house from "gadgetry."

The form of the berm creates an intimate courtyard, complete with tree (an oak), pool and terrace. The bank of solar panels is shaped into a gentle ellipse, to provide a more efficient collection surface. Under these panels lies a room with the tank storage and pumps.

A deliberate effort was made to clear the dwelling structure from the solar collection system.

In plan, the house is a "Modified American Ranch." In layout, bedrooms and baths lie to the east and family and dining rooms to the southwest. A garage, workshop and laundry are on a lower level, below the bedrooms.

The focus of the main rooms of the house is the courtyard, with its pool and terrace. It is imagined that the inner slopes of the berm might be planted entirely with flowers for a great summer yard.

Earth form and home are designed as one in an attempt to symbolize a natural house.

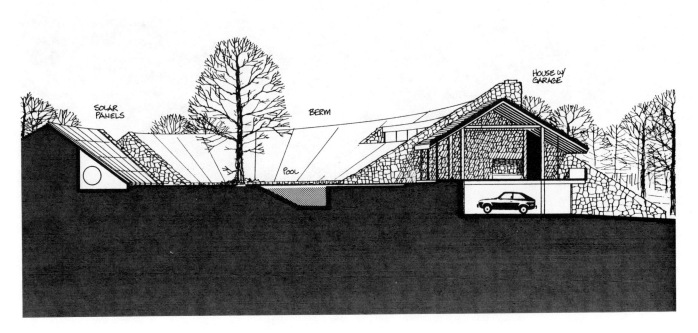

51

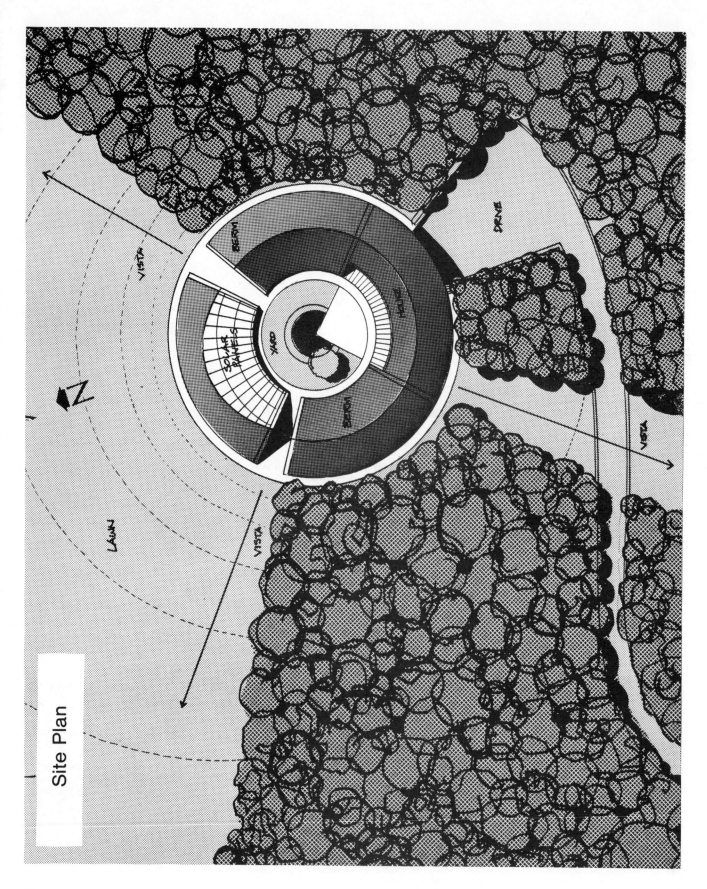

Site Plan

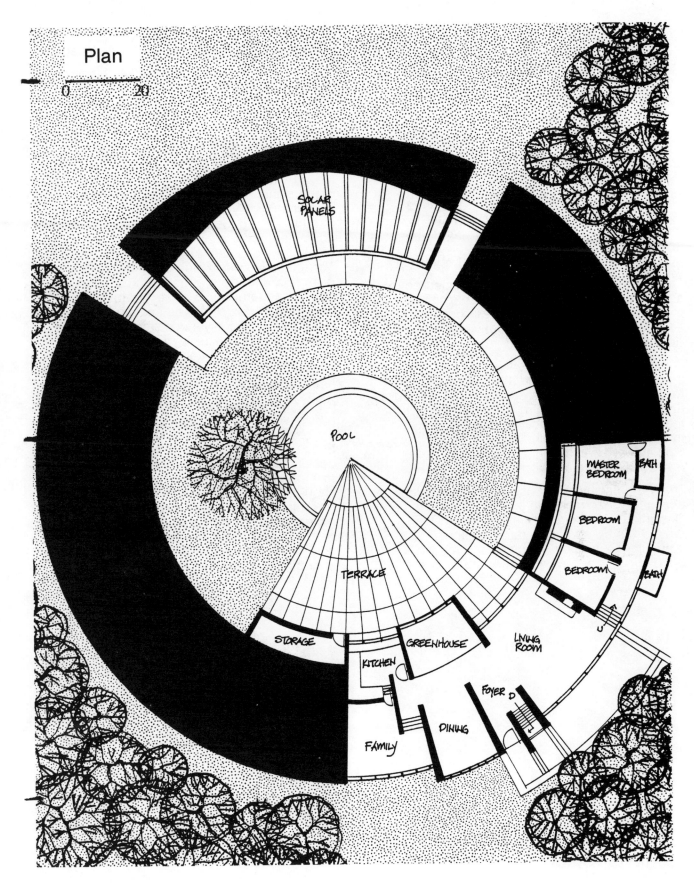

Plan

0    20

SOLAR PANELS

POOL

TERRACE

MASTER BEDROOM

BATH

BEDROOM

BEDROOM

BATH

STORAGE

GREENHOUSE

LIVING ROOM

KITCHEN

FOYER

DINING

FAMILY

## Vacation Home on Bay Lake, Minnesota by Milo H. Thompson, Frederick Bentz/Milo Thompson & Associates, Inc., Minneapolis, Minnesota

The Watson House, designed for a couple, two children and weekend guests, is a year round retreat house two and a half hours from Minneapolis on Bay Lake, Deerwood, Minnesota.

The site is a two acre peninsula characterized by a great mound rising from the water to an elevation of 36 feet. It is heavily wooded at the shoreline, and has very attractive views in all directions.

The owners indicated a preference for a plan arrangement which offered discreetly separate rooms that could provide privacy within the house and the potential of closing off unused facilities in the winter to minimize heating. They also wished that the house sit unobtrusively on the point of land and at the same time be a dramatic and eventful architectural statement.

The design solution consists of a tree-like plan configuration formed with a below grade skylighted trunk corridor running parallel to the crest of the hill and secondary corridors branching to individual structures which project out of the hillside. The structures are expressed as individual houses, each ending with a terrace formed from the excess excavation material.

Organic imagery has been attempted in the design — a tree structure, a leaf structure and a spatial sequence. Its construction process imitates an animal burrow. Rough timber and board finish for the building suggest that the structure is a natural part of the site.

Notions about energy and ecology are central ideas to this project. Earth-sheltered construction reduces the exposure of the building to the elements. Highly insulated construction is used throughout with the walls structured of reinforced concrete formed with molded hollow polystyrene blocks. The blocks are left in place as insulation. The roofs are constructed using heavy timber beams, one for each house, to create a valley of an inverted form which is pitched to drain at the back of each structure. The heating system is extremely simple, relying on small wood stoves, one for each house, using wood produced on the site. Only bathrooms and kitchen have back up electric heating and will be used throughout the heating season.

Currently in the working drawings stage, the house will be built starting in Spring, 1981. From the beginning, the owner and the designer have been committed to the idea of earth-sheltered construction. At the same time, however, the idea that technical and engineering expressions should dominate the design was rejected as too predictable, incomplete, and inappropriate. The owner and the architect intend this design to be a demonstration that earth-sheltered, energy-conserving design can be an artful combination of architecture and engineering.

**South Elevation**

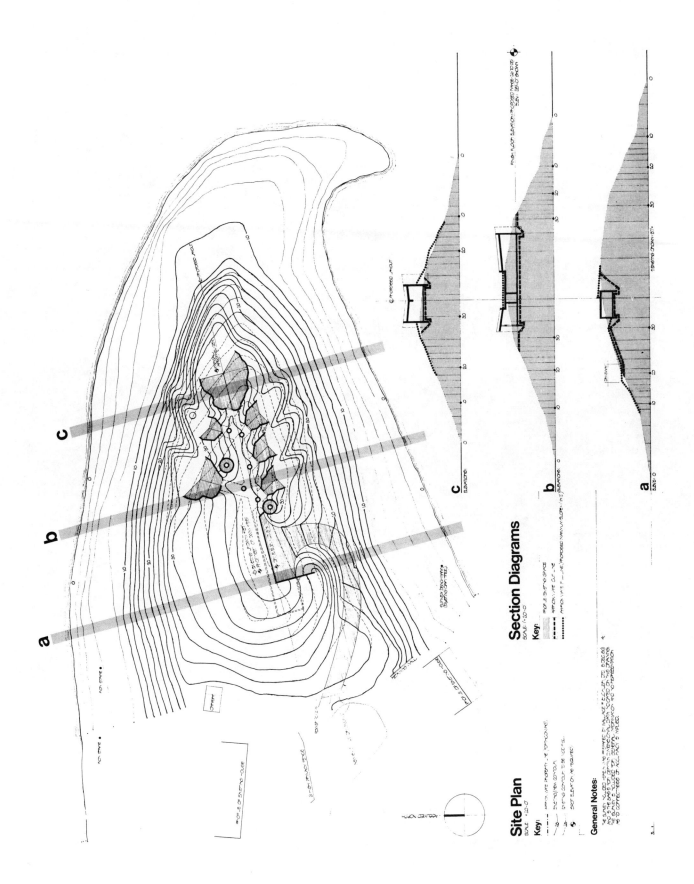

## Site Plan
SCALE 1:20'-0"

Key:

## Section Diagrams
SCALE 1:20'-0"

Key:

General Notes:

55

**BUILDING FORM** 02468 16 N

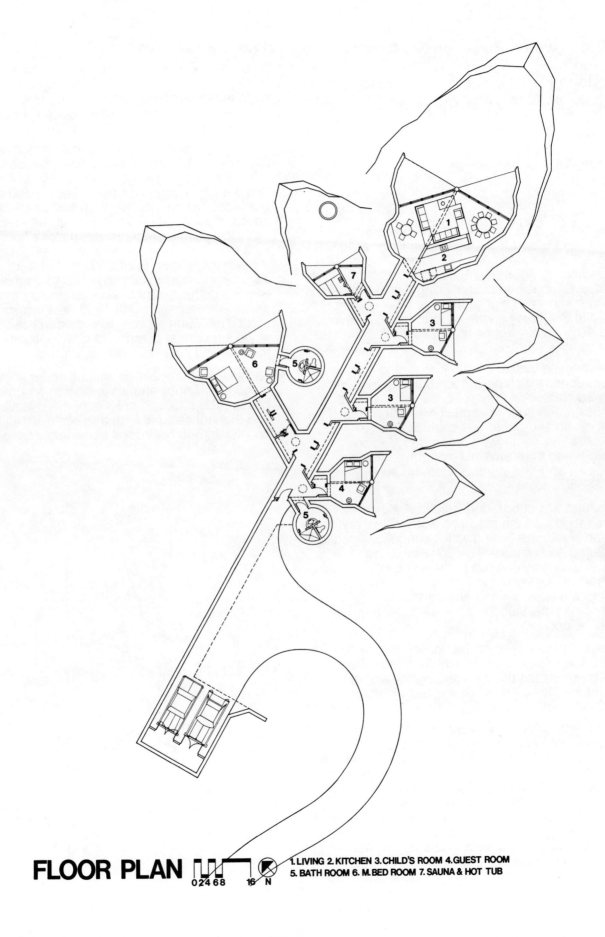

# FLOOR PLAN

0 2 4 6 8    16    N

1. LIVING 2. KITCHEN 3. CHILD'S ROOM 4. GUEST ROOM
5. BATH ROOM 6. M. BED ROOM 7. SAUNA & HOT TUB

# Wind Powered House on Nantucket Island
## by John H. Crowther, AIA, Architect, Montclair, New Jersey

The site is a 13.5 acre tract at the northwest tip of Nantucket Island, known as North Point, a sparsely populated part of the island. The vegetation is vigorous, varied and for the most part low. The terrain is undulating — mounds and gullies, known as glacial moraine. At one part of the site the moraine forms a rounded bluff which moves gradually from a north to a west orientation.

The moraine bluff to the north slopes to a series of sand dunes which have been building for years toward Nantucket Sound. To the west the bluff slopes to Eel Point, a large conservation area of salt marsh and dunes — hundreds of acres with no construction allowed. Beyond is Madaket Harbor and three small islands. On a clear day, you can also see Martha's Vineyard and Chappaquiddick.

The tract was subdivided into three water view lots, two lots on the road with lesser views and a lot on Nantucket Sound which cannot be used for building and which will be donated to the Nantucket Conservation Society. The lot lines are described legally but will not be defined by hedges, fences or other artificial divisions. The intent is to maintain the appearance of a unified tract of land.

A proposed single-family residence is located on lot 5. The design is an energy conscious design which meets the program requirements of the client and complements the island's historic architecture.

The bluff on this lot faces north to Nantucket Sound. The house is angled slightly to the northwest which allows for a more interesting diagonal view of the shore line and for better enjoyment of the magnificent sunsets. The siting also considers the south elevation which is designed with large glass areas to collect sunlight and a steeply pitched roof to accommodate future active solar collectors. The lower level is earth-sheltered to conserve energy and to reduce the structure's mass. The lower level bedrooms face south to a sunken court.

To conserve energy, the house is designed to be used partially or fully. Thus, the windmill portion is an independent living unit which can be occupied while the rest of the house is closed off and vice-versa.

Additional energy conserving and producing features are: a wind-powered, electrical generation system which will provide enough power for this and the adjoining residence; a load-shedding electrical system; passive solar system including solar collectors (south facing glass), an absorber and storage mass (concrete and stone floors and rock storage bin below the lower level slab); heat regulation devices (overhang and trellis for summer months, insulating blinds for nights and winter months, skylights and louvers to exhaust warm air in summer); a fan and duct system to recirculate warm air from the cathedral ceiling; wood stove; heat pump; and heavy insulation of the exposed areas of the structure.

The exterior materials are natural cedar shingles, masonry chimney and white dacron windmill blades. The subdued color scheme, materials, and the architectural form and scale are all consistent with Nantucket's building tradition.

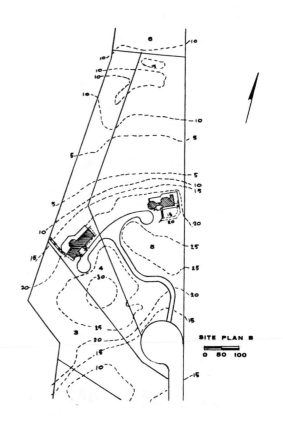

SITE PLAN B

0    50    100

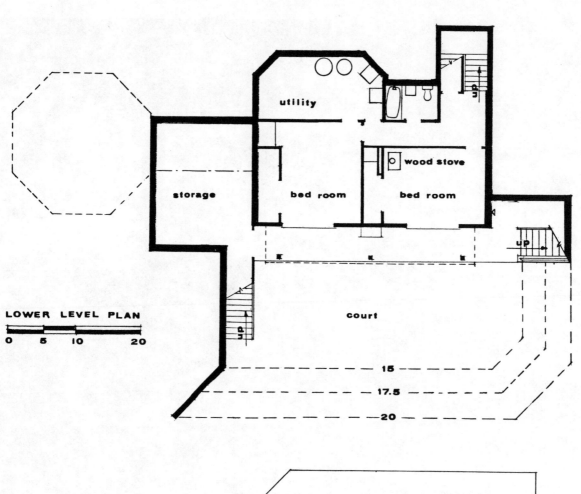

utility

wood stove

storage

bed room

bed room

LOWER LEVEL PLAN

0    5    10    20

court

15

17.5

20

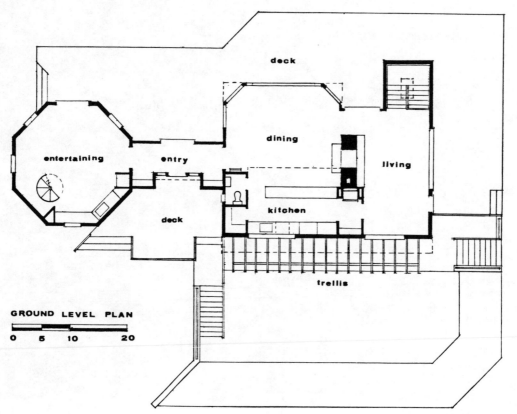

deck

entertaining

entry

dining

living

deck

kitchen

GROUND LEVEL PLAN

0    5    10    20

trellis

60

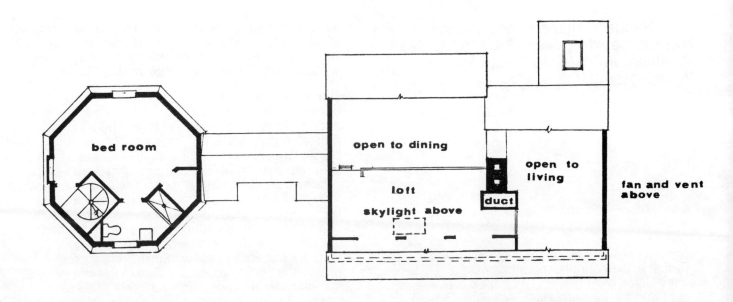

bed room

open to dining

loft

skylight above

open to living

duct

fan and vent above

SECOND LEVEL PLAN

0    5    10          20

NORTH ELEVATION

## Park Ranger's Home in Washington State by Prof. David M. Scott, FAIA, Asst. Dean, Washington State University, Pullman, Washington

A large dam was developed along a major western river. The scale of the river and canyon is grand and very rural — nearly a wilderness area. The pool of water extends for several miles. Together, the county park department and U.S. Army Corps of Engineers developed a plan for a park with the usual facilities — toilets, covered picnic area, recreational vehicle sites, boat launching facilities and a home for the park ranger.

State agencies funded a planning grant to develop a feasibility study and preliminary drawings to determine if an earth-sheltered structure might be appropriate as a ranger's home. The study concluded that the project shall be a demonstration house accessible to the public.

The architect, in consultation with the park department staff, the park board and the county commissioners, developed the plans and specifications. Support was received from county, state and federal institutions, local banks, utilities, suppliers, manufacturers, contractors, craftsmen, professionals and students. Construction was started and the project will be substantially complete by June of 1981.

During the early programming stage, the following goals were established: provide a house that will be safe and habitable for all ages, including those with impaired mobility; provide space for a family of four; demonstrate that a house can have minimum visual and ecological impact on the landscape; take advantage of passive solar gain and the tempering effects of the soil and demonstrate the energy efficiency of earth-sheltering; create a structure that has minimum exterior maintenance; develop a house that can be built with the same craft skills, using the same materials, products and components as traditional houses, while incorporating new materials; and develop a house that is full of light, warm and dry, creating a strong sense of security and place.

The primary emphasis in the project has been to develop a thermal efficient shell. The public side (ranger's office) contains the main entry and vehicle access. The private side (family) addresses the view and provides a private yard.

The house was developed as a cast-in-place concrete shell with insulation on the outside of the concrete. Insulation exposed to air and sun is protected by the "DRYVIT" system. Thermal breaks occur wherever structurally feasible. There is approximately 5,400 cubic feet of exposed concrete used as thermal mass. The glazed areas face southeast and southwest. The axis of the site is to the south. Because the house is at the mouth of a canyon, and a 600 foot hill partially shades the site during the months of December and January, access to view was given a higher priority than south exposure. Due to the intense summer sun, a shaded wall was generated as opposed to using either a trombe wall or a greenhouse concept.

The house has a counter-flow, air-to-air heat pump. An air-tight wood stove has been provided as backup. Time switches have been provided on all fans. Appliances were selected for energy conservation and cost. Sliding glass doors are double insulated glass with thermal breaks. Three

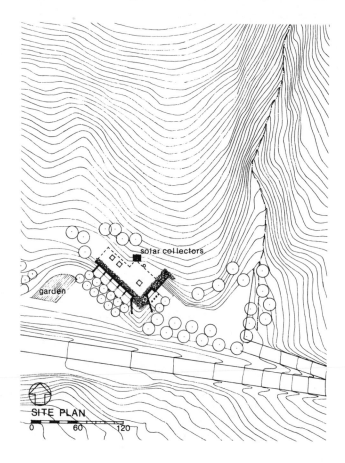

SITE PLAN
0   60   120

solar panels fitted into the contours of the hill preheat the domestic hot water.

There are two particularly unique construction features: a ceiling height of 9′-2″ is created on the living side of the house generated by the functional requirements of a ⅜ inch per foot slope required to drain water off the roof. And there is a 6′-8″ shelf or valance at the edge of most rooms. This valance visually controls the many openings and cabinets that relate to the shelf while providing a cove for indirect lighting and a shelf for the display of objects. Because of the removable valance top, there exists an accessible horizontal chase in which power, television, telephone and speaker cables are now housed. The chase provides easy access for future changes or additions.

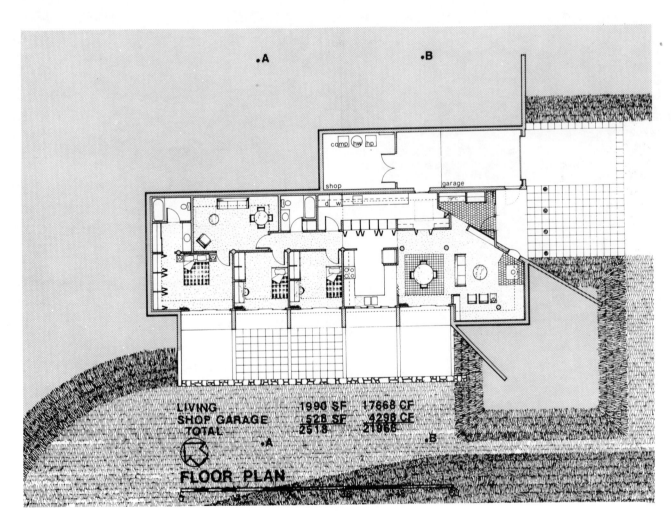

LIVING 1990 SF 17668 CF
SHOP GARAGE 528 SF 4298 CF
TOTAL 2518 21966

FLOOR PLAN

**THE PRIVATE SIDE**

THE PUBLIC SIDE

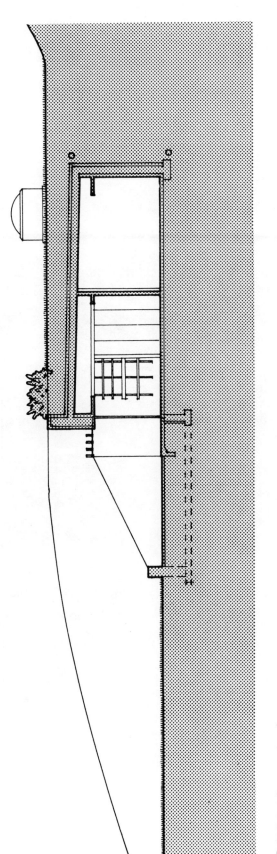

SECTION A•B

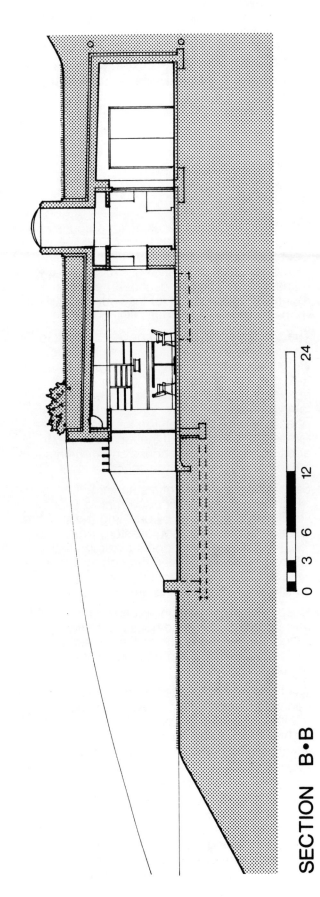

SECTION B•B

0 3 6 12 24

65

## Country Home in the Texas Hill Country
by Andrew Perez and John Grable,
Architects, O'Neill, Perez and Associates,
San Antonio, Texas

traditionally a massive form within a room, is treated as an exterior free-standing unit which corbels through a living room window.

The client requested a design for a country home on a five acre tract of land in the Texas Hill Country. The program includes a low-maintenance, energy-efficient structure that would respect the existing landscape. The site is essentially level except for the south end which drops off dramatically to a valley below.

The south sloping site is a perfect candidate for an earth-sheltered structure. By sinking the structure into the hillside, advantage is taken of the earth's great resistance to temperature change. At a depth of six feet, the earth's temperature is relatively constant at 70°F. This phenomenon helps stabilize the seasonal temperature extremes of the rooms above. In addition, the merger of house and hill minimizes the impact of a man-made structure to the site.

San Antonio's climate offers numerous days which do not require air conditioning. Cross ventilation is achieved by orienting the structure toward the prevailing southeasterly breeze. Eight foot high sliding glass doors maximize this air intake which is exhausted through two casement windows flanking the front door and mechanical ventilators at the dressing areas.

Deep overhangs over tinted and insulated sliding glass doors exclude the hot summer sun. The low-latitude winter sun is allowed to penetrate and radiant heat is stored in the concrete floors for nighttime use.

The house is constructed using standard low-cost concrete materials readily available and manufactured in the area. Both the client and the architect expressed concern about the austere feeling of these materials. This is solved by application of warm color stain to the concrete floors and by the use of regional historical details.

The confined feeling usually associated with earth-shelters is tempered with the use of nine foot ceilings and a south facing wall composed entirely of sliding glass doors. The element of surprise is achieved by entering through the front door on the opposite wall. The fireplace,

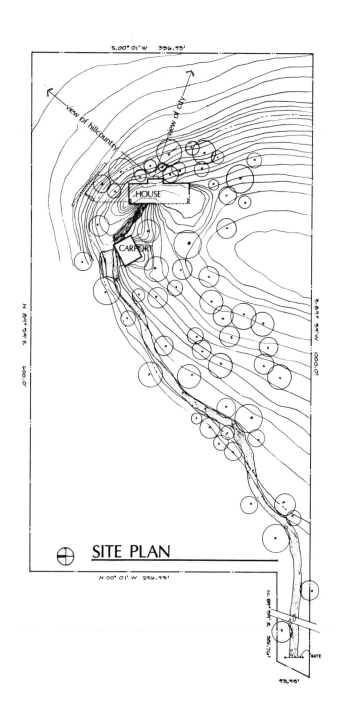

SITE PLAN

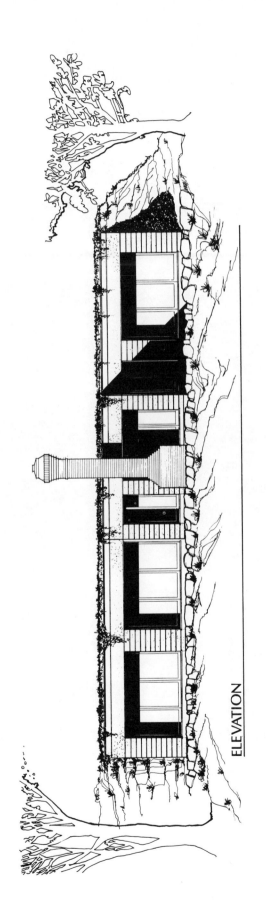

ELEVATION

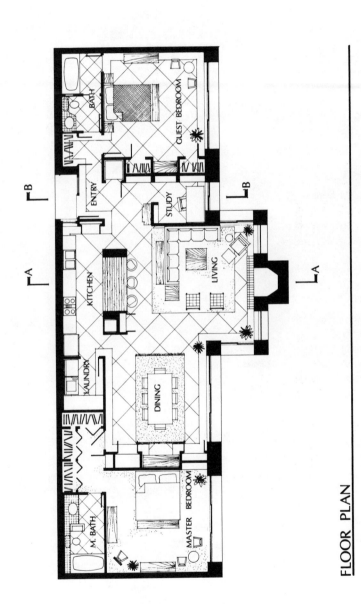

FLOOR PLAN

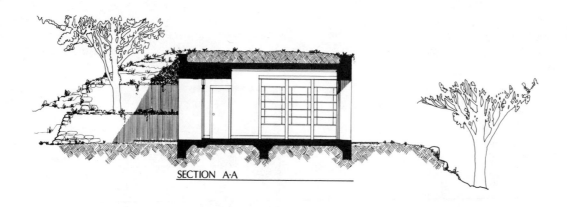

SECTION A·A

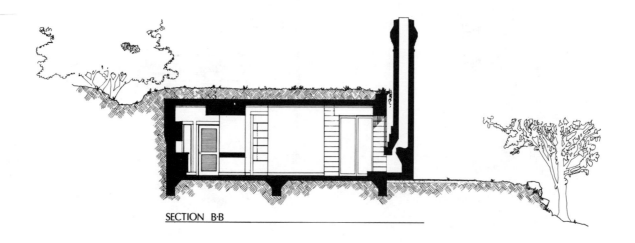

SECTION B·B

# Manager's Home on a Horse Farm in Kentucky's Blue Grass Region
by James B. Tune, President, J.B. Tune Associates, Architects, Lexington, Kentucky

The Blue Grass Region is a gently rolling plateau with elevations varying between 900 and 1,050 feet above sea level. The project is located 38° 02′ N Latitude. The climate is described as "temperate continental" with a rather wide daily temperature range, as there are no bodies of water close enough to provide any tempering effect. Prevailing winds are southerly in all seasons, but inclement winter weather is generally associated with westerly winds.

This residence is nestled in a southeast-facing slope. Because of the nature of the farm manager's work, the structure had to be located within the physical confines of the 140-acre farm, in close proximity to the farm's operations. The house's location takes advantage of the southerly summer breezes and provides shelter from the west.

The visual impact of this compact structure on the surrounding environment is minimal. Only an eastern and southern elevation are exposed; the latter supplies a magnificent view of the farm. The floor of the master bedroom is elevated ten inches higher than the rest of the house, contributing to a sense of privacy. A clerestory above the alcove allows daylight to penetrate the study and corridor and adjacent spaces. With the exception of the bathrooms, storage and mechanical rooms, all spaces have windows and operable vents providing a view, natural lighting and ventilation. All closets and other dead air spaces are kept free of the retaining wall to eliminate mildew, a considerable problem in the humid Kentucky climate.

Foundation walls and roof slab are reinforced concrete, interior finishes are painted concrete and rough-sawn cedar, and flooring is oak over a concrete slab on grade. Bentonite panels provide waterproofing on both horizontal and vertical surfaces. Selected to minimize maintenance, exterior finishes are unpainted concrete, rough-sawn cedar and treated wood at the trellises. Insulation in contact with earth is rigid polystyrene. All other is fiberglass batt.

A wood-burning stove with outside air intake will be the principal source of heat. South-facing, double-insulated glass will allow for heat gain in the winter, with plant-covered trellises providing shade in the summer. In addition, cove-mounted return-air grilles will capture heat from lights and stratified hot air at the ceiling for filtering and redistribution. All glass areas will be supplied with manually-operated thermal shutters, and a heat pump will serve as the back-up system. Concrete walls and piers provide a large thermal mass allowing floors to be wood for psychological and physical comfort.

Earth-sheltering was used primarily to diminish the visual impact of the building on the site and to avoid significantly disturbing the rest of the farm surroundings. Other structures in proximity are "farm" buildings specifically designed to accommodate thoroughbred race horses. The energy benefits associated with earth-sheltering were originally secondary to the site consideration. However, design development included the integration of many other energy-saving concepts, such as passive solar adaptations. Heavy wood trellises relate the building to the main residence and to wood framing details on other farm buildings.

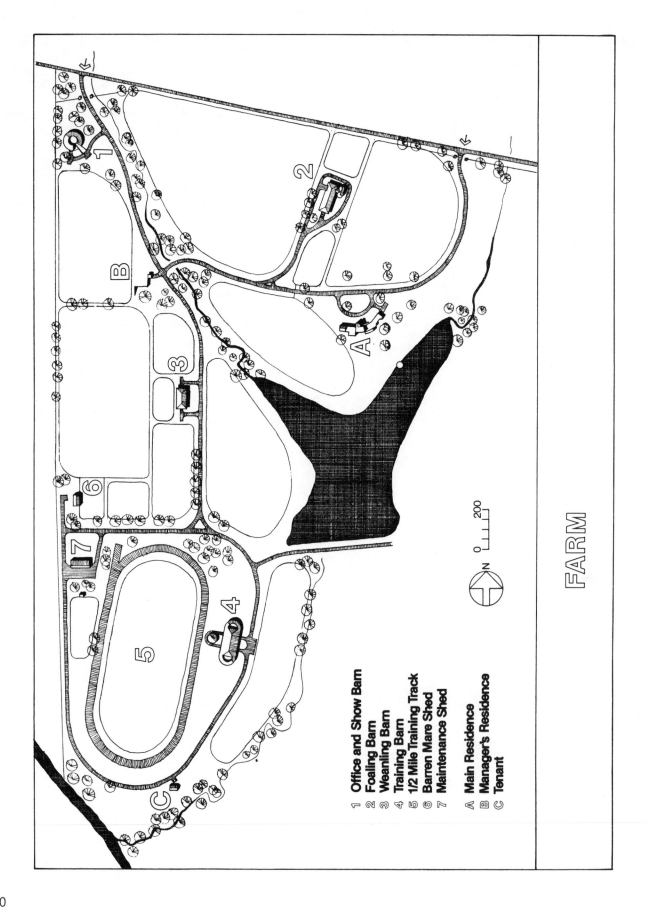

1 Office and Show Barn
2 Foaling Barn
3 Weanling Barn
4 Training Barn
5 1/2 Mile Training Track
6 Barren Mare Shed
7 Maintenance Shed

A Main Residence
B Manager's Residence
C Tenant

N 0 200

FARM

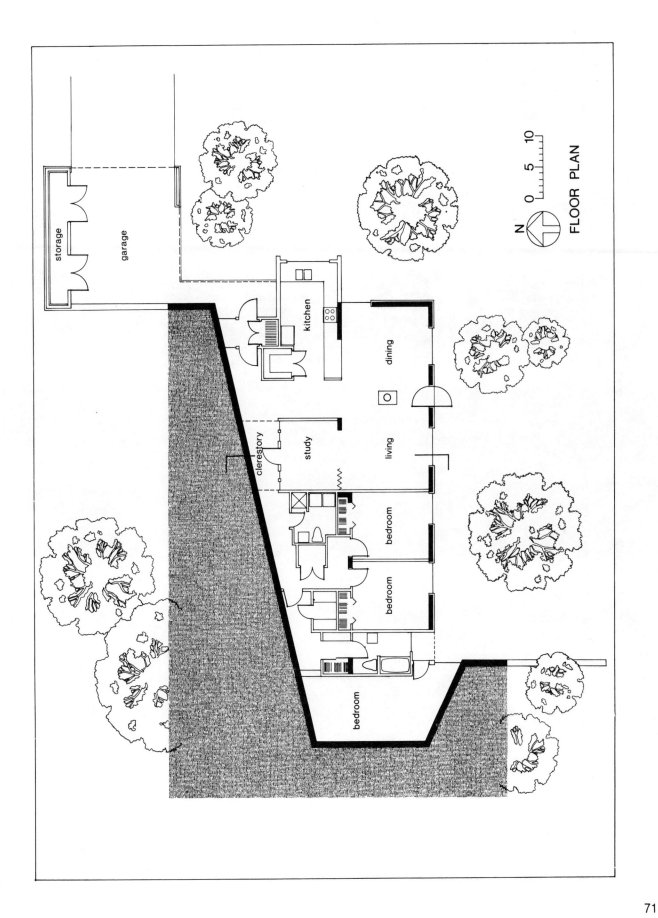

storage

garage

kitchen

dining

clerestory

study

living

bedroom

bedroom

bedroom

N

0 5 10

FLOOR PLAN

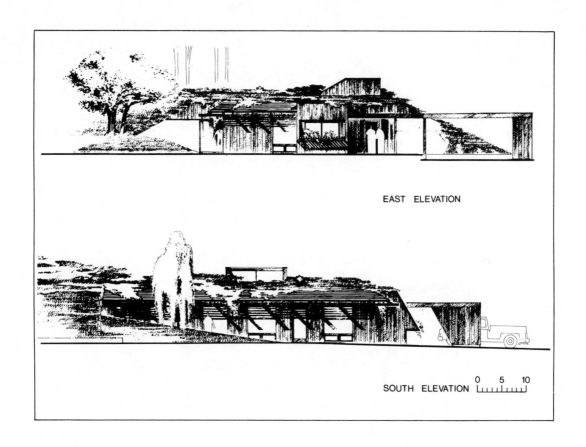

EAST ELEVATION

SOUTH ELEVATION    0   5   10

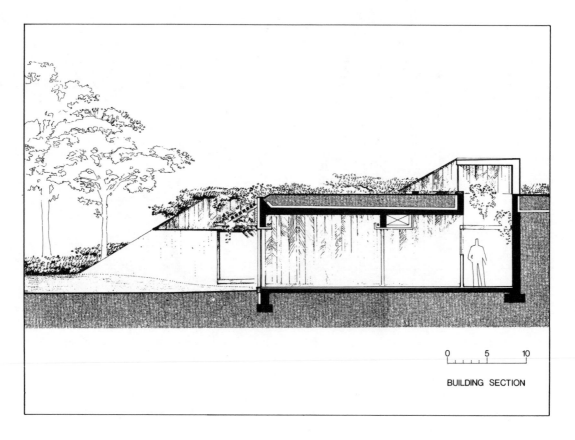

0    5    10

BUILDING SECTION

## Home near Mangum, Oklahoma
by Joe Mashburn, Architect, College Station, Texas

Residence for Dr. and Mrs. Jeffry S. Lester and family, Mangum, Oklahoma.

The following are quotes from client's letters: "Our view to the east is a mountain range about eight miles away and (there are) no visible roads. We would like (the house) built into the side of a hill. We would like it facing east, but still (to) have efficient passive solar. . . . There are no general contractors nearby. . . . Basically, we want quality living, low upkeep, and unusual architectural design."

The proposed solution includes several concepts: the west greenhouse acts as a winter solar collector; a breezeway isolates the greenhouse

from house in summer. Overhangs are calculated for location. Interior stone and depth are designed to conserve winter heat. Natural ventilation is aided by the orientation of the breezeway, ravine and exterior stone walls. Protected entries are provided.

The house is sited to take advantage of view across the property (fields of alfalfa) to the granite mountains beyond.

Proposed materials include: precast concrete, double "Ts" on precast beams and concrete walls; exposed steel columns; native stone walls at exterior and lower interior; native stone and carpet floors; gypsum board walls at the upper level; precast is used for stability and in order to subcontract much of the erection to an Oklahoma City concrete company for rail and truck delivery. All glass is operable. Insulated drapes are used in most areas.

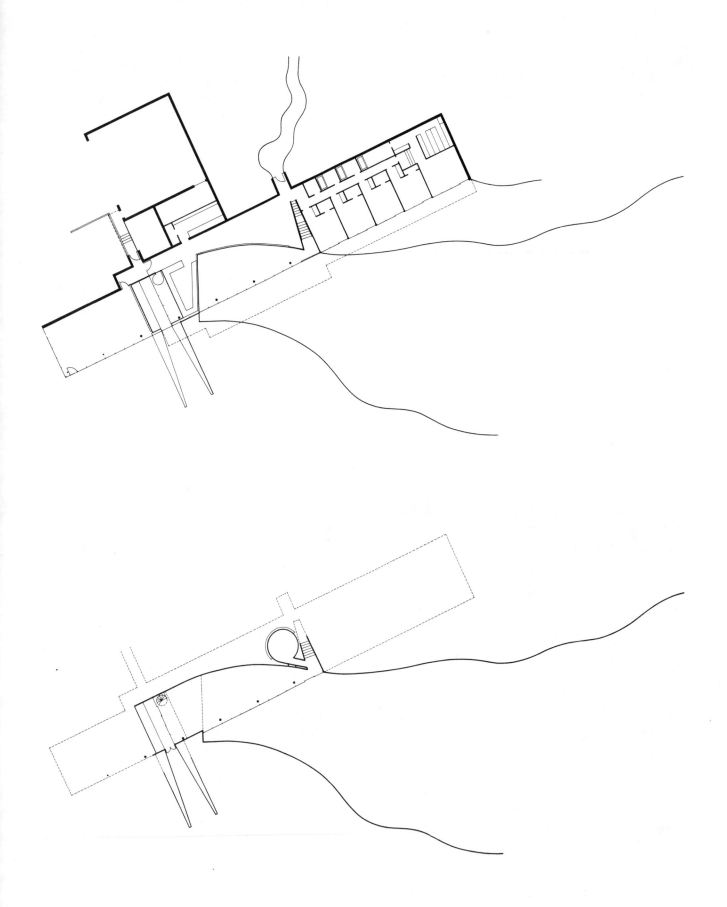

# Home with a view near Colorado Springs, Colorado
## by Ted R. Locke, AIA, Architect, Colorado Springs, Colorado

and the city of Colorado Springs can be seen to the south from both interior and exterior spaces. The interior spaces are zoned for entertainment (upper level) and family use (lower level). Energy conservation and low exterior maintenance are additional considerations which influenced this earth-sheltered design solution.

The site is located on the Monument Ridge, north of Colorado Springs, Colorado. Elevation of the site is approximately 7,000 feet. Frequent, fast moving wnter storms from the northwest are characteristic of the area with wide temperature variations occurring within a short time span. The site slopes to the southeast. A vertical difference of 42 feet exists from the high to low point. The vegetation consists of scrub oak, ponderosa pine and native grasses.

The residence is designed for a family of four with strong interests in music and art. Outdoor terrace areas are required for entertaining clients and friends. The U.S. Air Force Academy, Pikes Peak

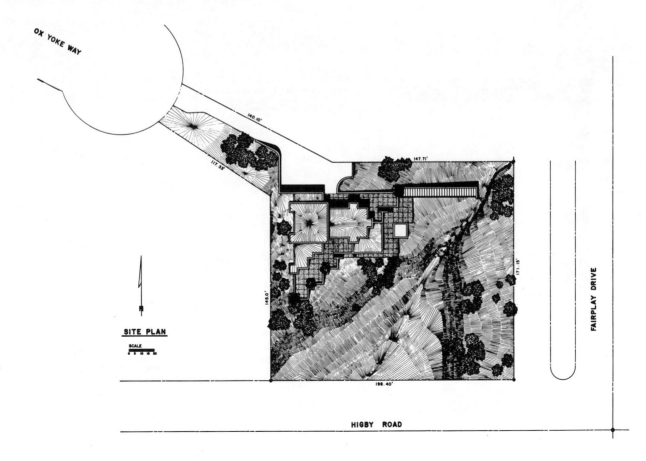

OX YOKE WAY

SITE PLAN

SCALE

FAIRPLAY DRIVE

HIGBY ROAD

75

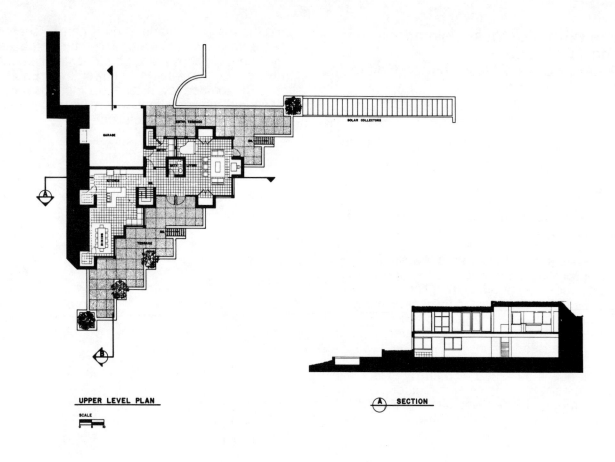

**UPPER LEVEL PLAN**

SCALE

(A) **SECTION**

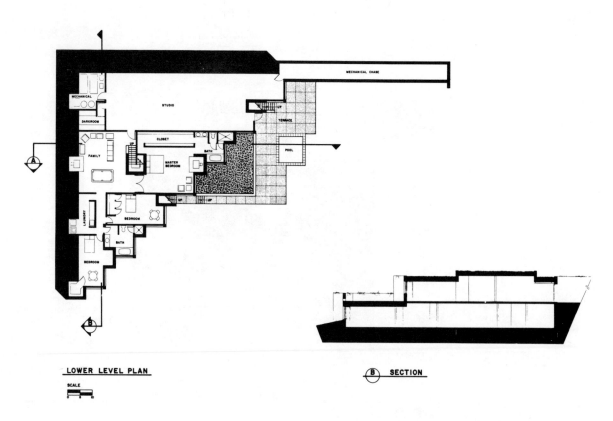

**LOWER LEVEL PLAN**

SCALE

(B) **SECTION**

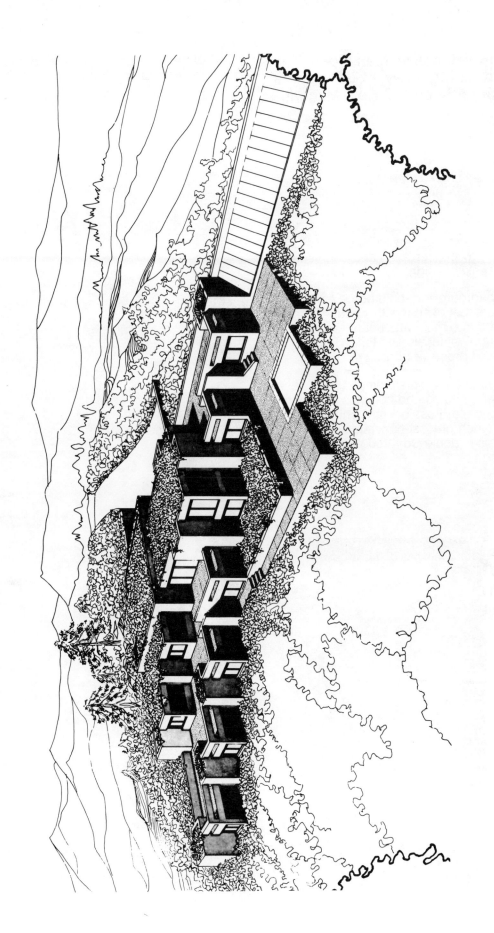

## Circular Residence with an Open Plan
by Robert M. McCulley, AIA, President, Wolfenbarger and McCulley, Manhattan, Kansas

of the insulative properties of the earth and by natural lighting from the centrally located courtyard.

This project combines open planning, natural lighting, energy conservation and earth-sheltering into a residential design.

The three-bedroom, four-bath home encloses 3,400 square feet of space. Its circular shape encloses the maximum amount of area with minimal use of materials and tends to resist horizontal soil pressure forces.

The basic structural system employed is concrete bearing walls, beams and columns supporting precast concrete "truss-beams." The roof is made up of eight inch precast concrete slabs. Energy conservation is achieved through maximum use

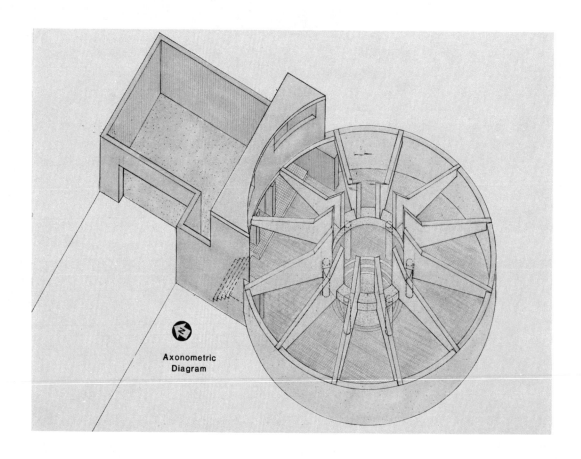

Axonometric
Diagram

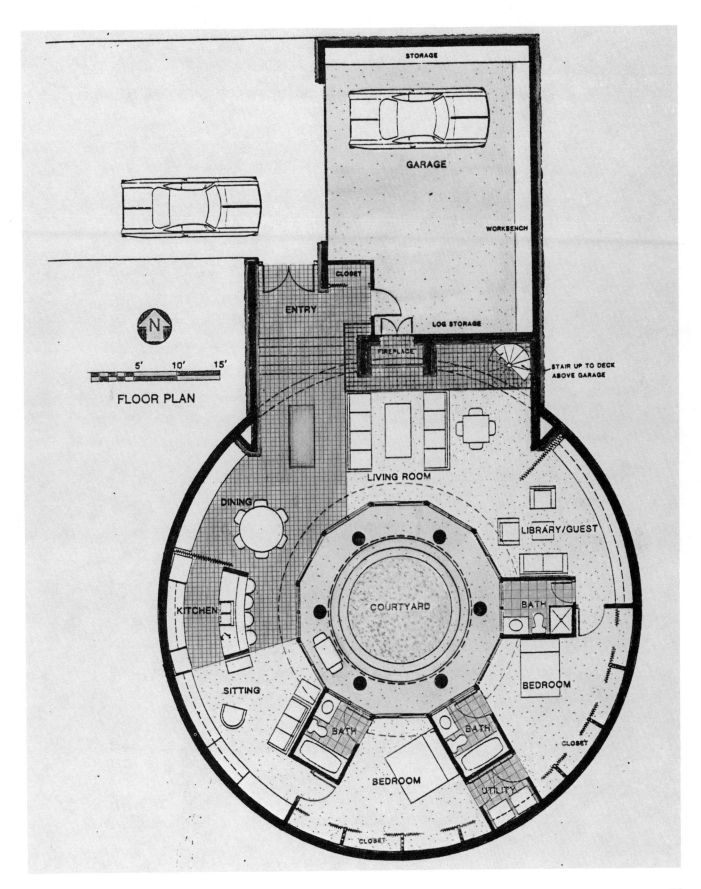

STORAGE

GARAGE

WORKBENCH

N

5'  10'  15'

FLOOR PLAN

CLOSET

ENTRY

LOG STORAGE

FIREPLACE

STAIR UP TO DECK
ABOVE GARAGE

LIVING ROOM

DINING

LIBRARY/GUEST

COURTYARD

BATH

KITCHEN

BEDROOM

SITTING

BATH

BATH

CLOSET

BEDROOM

UTILITY

CLOSET

79

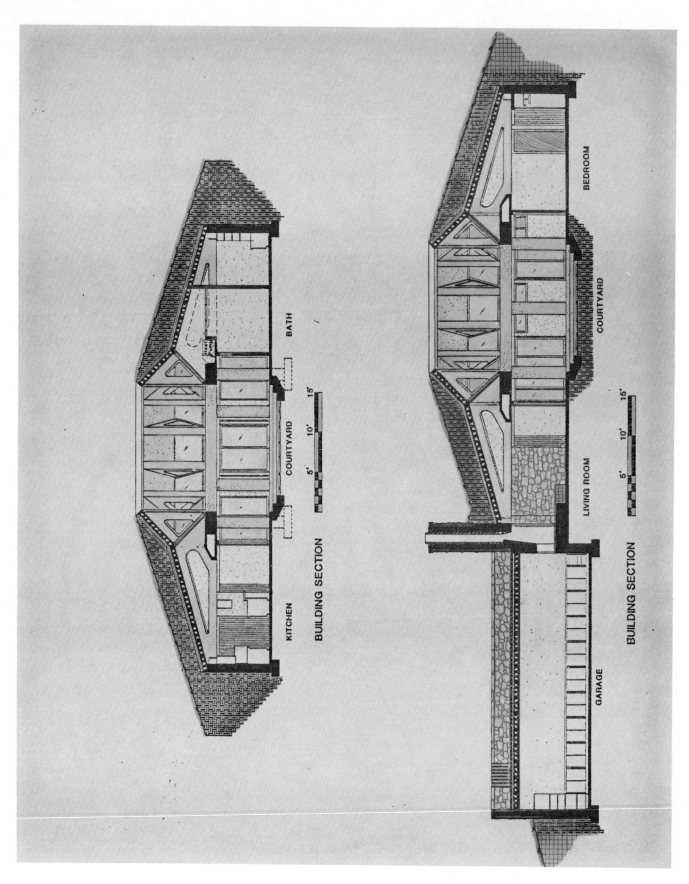

BATH

COURTYARD

KITCHEN

5'    10'    15'

BUILDING SECTION

BEDROOM

COURTYARD

LIVING ROOM

GARAGE

5'    10'    15'

BUILDING SECTION

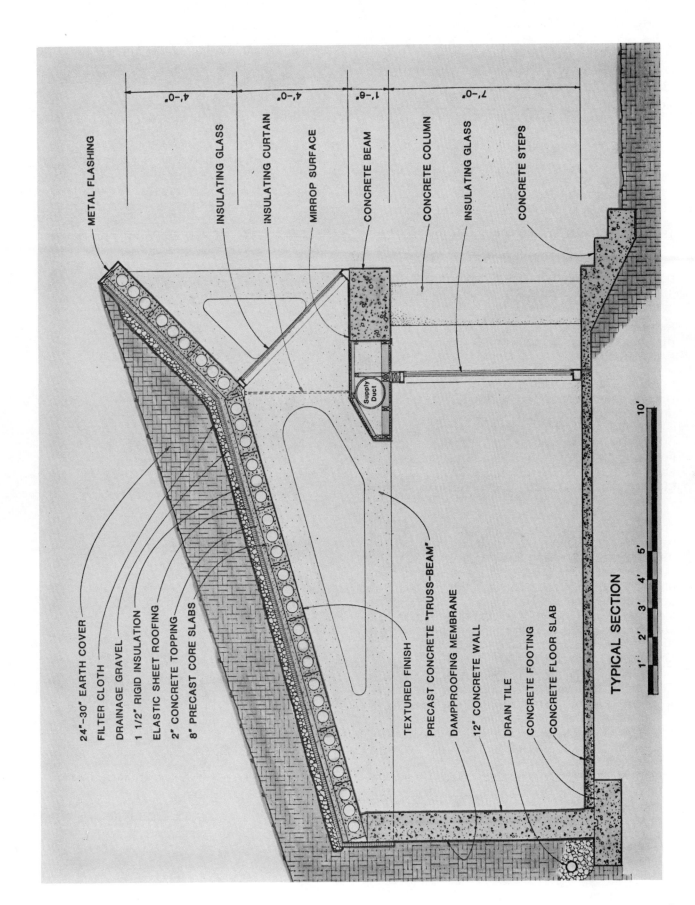

TYPICAL SECTION

4'-0"  4'-0"  1'-6"  7'-0"

METAL FLASHING
INSULATING GLASS
INSULATING CURTAIN
MIRROR SURFACE
CONCRETE BEAM
CONCRETE COLUMN
INSULATING GLASS
CONCRETE STEPS

24"-30" EARTH COVER
FILTER CLOTH
DRAINAGE GRAVEL
1 1/2" RIGID INSULATION
ELASTIC SHEET ROOFING
2" CONCRETE TOPPING
8" PRECAST CORE SLABS

TEXTURED FINISH
PRECAST CONCRETE "TRUSS-BEAM"
DAMPPROOFING MEMBRANE
12" CONCRETE WALL
DRAIN TILE
CONCRETE FOOTING
CONCRETE FLOOR SLAB

Supply Duct

1'  2'  3'  4'  5'    10'

81

## A Seasonally Adaptive House
by Willis P. Lawrie, Designer, Sunhouse Division, R.L. Light Architect, Bristol, Tennessee

The emphasis and theme of the design is the total integration with the natural environment and with the cycles of nature. Like a sailing yacht, the house is composed of a 'selective hull' or earth-integrated structure and a 'variable rigging' or energy-tempering system.

Sited in a rolling green meadow at the south edge of a small woods, the house is visually linked to its surroundings.

In general, the more internal environment of winter is segregated from the external natural environment through layers of space. In summer, a continuous progression from the inside to the outside creates 'gradients' in temperature and air flow. Varying the enclosure at the western and southern edges enables changes in the indoor-outdoor relationships during the course of the year.

The west elevation provides the major entries as well as a porch in summer. The glass wall can be moved aside integrating this space with the outdoor space. The outdoor space is shaded from afternoon sun by trees. Other functions of this space are the winter storage of four to six cords of firewood as well as the collection of afternoon sunlight. Since it is unheated, this space maintains a median climate between the internal and external climates.

The southern elevation, enclosed by a glass roof in winter, is more dramatically attuned to seasonal variations and the adjustments. In addition to being used as a solarium or greenhouse during the winter, the kitchen, dining room, living room and bedrooms can open and extend into this space through a movable glass wall. In winter, the transparency of the envelope is controlled by insulated louvers which follow the motion of the sun. Heat is absorbed by the masonry walls and water-filled drums. Natural convective air flows, added by fans which are located at the top of the thermal storage mass, provide continuous circulation. On sunless days, the glass partitions remain closed and air circulation is maintained only in the internal spaces. Also during sunless days, the wood burning system becomes the primary energy system.

During spring, adjustments can begin to be made for warmer weather. The louvers can be moved into a fixed summer position to provide shade for the southern elevation. The movable walls at the south and west can be opened to allow natural ventilation through the house. The dome above the stair may be opened to the north, providing an outlet for the prevailing breezes. Shade may also be supplied by vines growing from the sod and earthen roof covering.

**View    from    southwest**

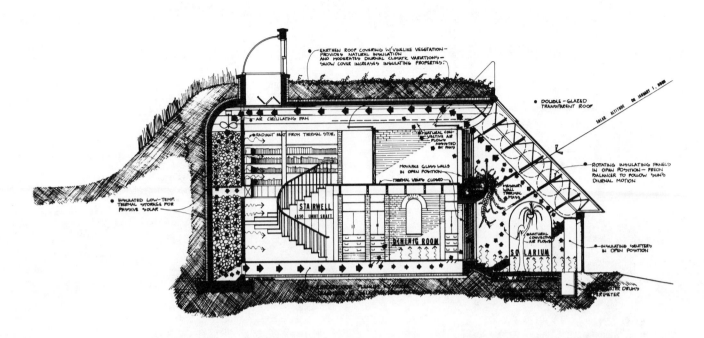

## NORTH~SOUTH SECTION: WINTER
(SHOWING SUNNY DAY OPERATION)

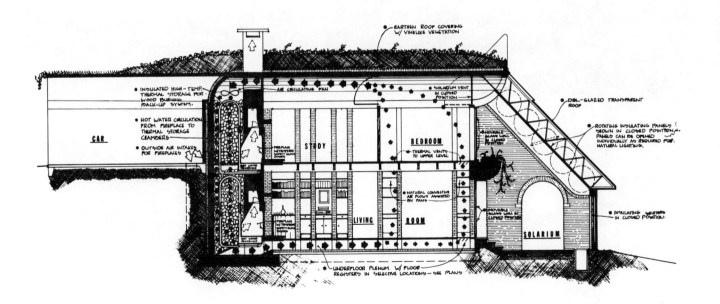

## NORTH~SOUTH SECTION: WINTER
(SHOWING CLOUDY DAY & NIGHTIME OPERATION W/ WOOD-BURNING
BACKUP SYSTEM)

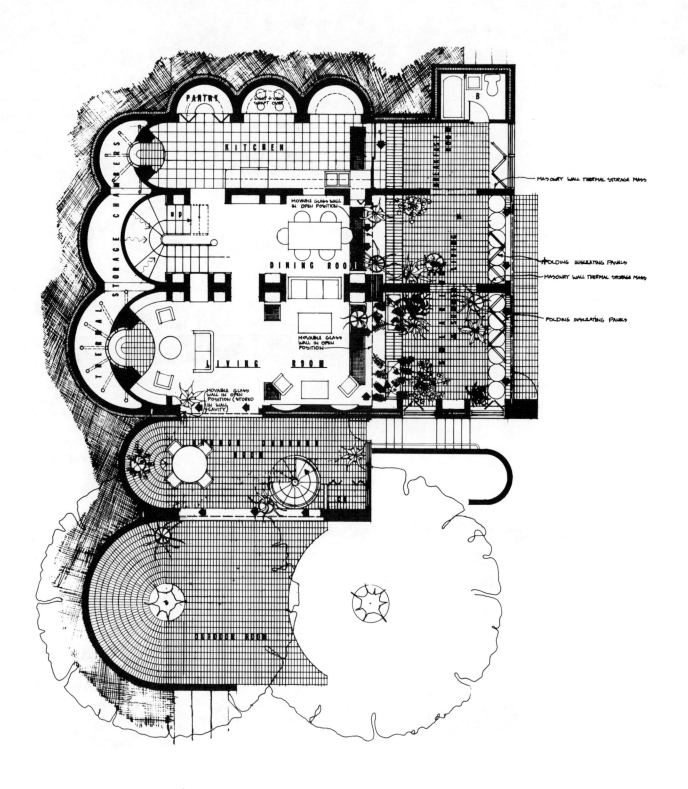

MASONRY WALL THERMAL STORAGE MASS

FOLDING INSULATING PANELS

MASONRY WALL THERMAL STORAGE MASS

FOLDING INSULATING PANELS

PANTRY

LIGHT + VENT SHAFT OVER

B

KITCHEN

THERMAL STORAGE CHAMBERS

UP

MOVABLE GLASS WALL IN OPEN POSITION

BREAKFAST ROOM

DINING ROOM

MOVABLE GLASS WALL IN OPEN POSITION

LIVING ROOM

MOVABLE GLASS WALL IN OPEN POSITION (STORED IN WALL CAVITY)

INDOOR ROOM

LOWER LEVEL PLAN

0 1 2 3 4 5          10

NORTH

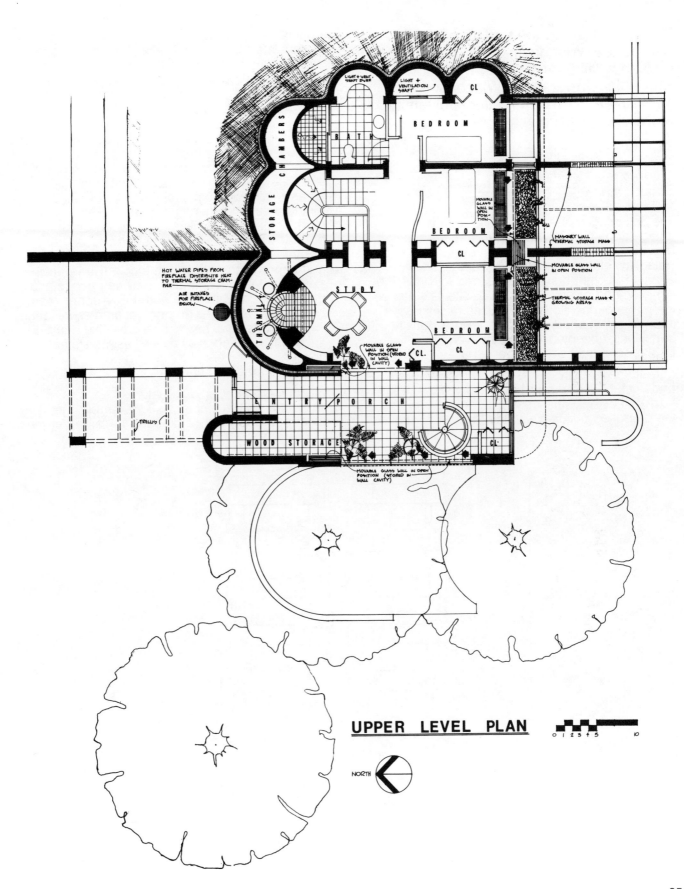

LIGHT & VENT
SHAFT DUCT

LIGHT &
VENTILATION
SHAFT

CL

BEDROOM

STORAGE CHAMBERS

BATH

MOVABLE
GLASS
WALL IN
OPEN
POSITION

MASONRY WALL
THERMAL STORAGE MASS

BEDROOM

CL

MOVABLE GLASS WALL
IN OPEN POSITION

HOT WATER PIPES FROM
FIREPLACE DISTRIBUTE HEAT
TO THERMAL STORAGE CHAM-
BER

THERMAL

STUDY

THERMAL STORAGE MASS &
GROWING AREAS

AIR INTAKES
FOR FIREPLACE
BELOW

BEDROOM

CL

MOVABLE GLASS
WALL IN OPEN
POSITION (STORED
IN WALL
CAVITY)

CL

ENTRY/PORCH

TRELLIS

WOOD STORAGE

CL

MOVABLE GLASS WALL IN OPEN
POSITION (STORED IN
WALL CAVITY)

## UPPER LEVEL PLAN

0 1 2 3 4 5          10

NORTH

## Proposed Residence for Urban Milwaukee, Wisconsin
by Nanci Miller and Karen Markison, Students, University of Wisconsin-Milwaukee/Carnegie Mellon University, Milwaukee, Wisconsin

The advantages of earth-sheltered buildings in energy and environmental terms already seem clear. This design represents an exploration of the potentials of earth-sheltered buildings for enhancing other human needs, and for providing richness and diversity. The urban site selected on the fringe of downtown Milwaukee raises issues such as street character, scale, texture, density, transitions, acceptability and the hierarchy of public, semi-public and private spaces.

Milwaukee is a city of one- and two-story wood and brick houses which have a very strong front/back, street/alley relationship. Efforts to integrate the proposed building with the Milwaukee urban context resulted in a design using scale and materials similar to neighboring structures. The structure's relationship to the street, its facade and landscaping elements help put the building in proper context. The wall also provides a strong sense of transition from public to private spaces.

The configuration of each house and the block as a whole creates hierarchies of spaces in three dimensions. The east-west axis on the ground level is a graduation of public to semi-public areas — moving from street to entry to living/dining area to kitchen to garage and alley. The north-south axis consists of a private space bounded on both sides by public spaces. Commercial and community functions lie on the north and south ends of the block. The vertical axis provides for separation between public and private spaces within the home. The upper, more public level assumes an open character while the lower, private level has almost cave-like quality. These intersecting axes as well as the stair element give a central focus to a house which is otherwise long and narrow.

Within the house, energy features become important design elements. For example, the greenhouse is a habitable transition area between

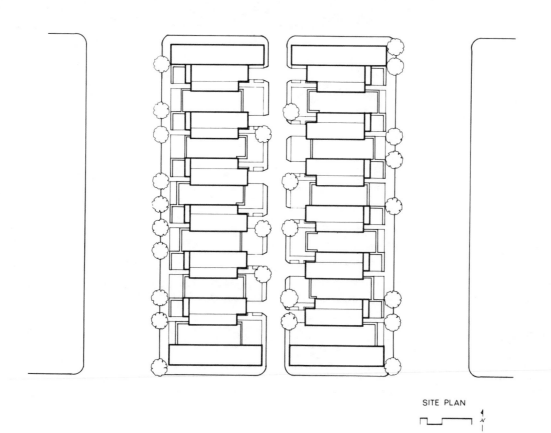

SITE PLAN

the open courtyard and the more closed home.

Passive solar energy is the major heating source. An active solar system with freon panels is used to heat water. Other energy sources are provided as backup. Air is brought into the house via a rock storage bed buried in the yard. In summer, ventilation is provided through a clerestory.

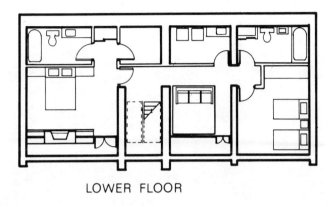

LOWER FLOOR

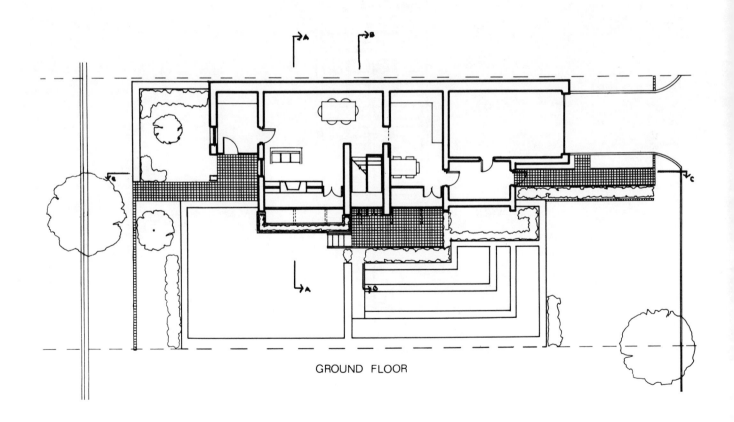

GROUND FLOOR

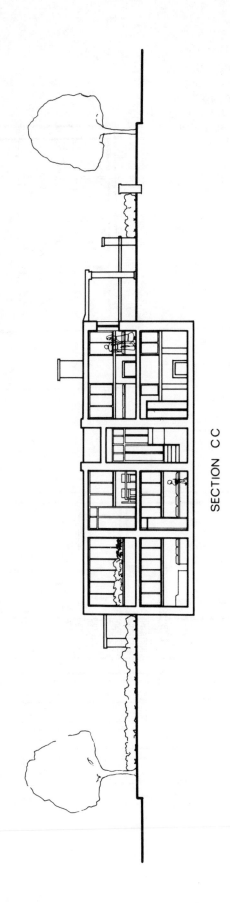

SECTION CC

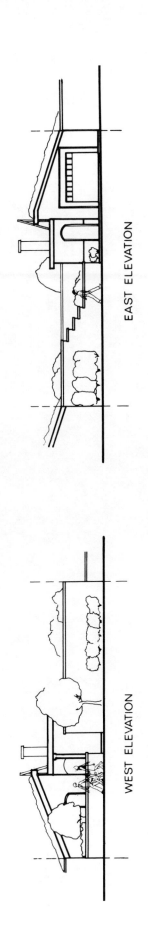

EAST ELEVATION

WEST ELEVATION

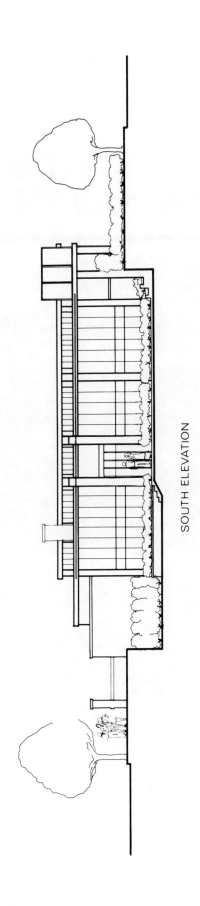

SOUTH ELEVATION

# Multi-Family Residential

"I believe, beyond any doubt, that Multi-Family Residential is the most difficult category to do in an underground earth-sheltered situation. As an undergraduate, I tried to design housing which stepped up a hill. I discovered very quickly that there are real problems with cross-ventilation, access to view and privacy.

"Most of the entries are very much like double-loaded corridor apartments except that there is earth where the double-loaded corridor would be and there is a conflict between view and privacy where an apartment normally only has a view.

"There are not that many good housing projects, let alone good earth-sheltered housing. Designers of earth-sheltered projects should look at the precedence of row housing in densely urban areas where a stoop or a change in grade provides privacy, access to views and entrances in limited space without conflicts."

Edward Allen

# Merit Award

## Residential Complex for Minnesota by Milo H. Thompson, AIA, Frederick Bentz/Milo Thompson and Associates, Inc., Minneapolis, Minnesota

A study was conducted to show city officials and the owner of the site, a concrete block manufacturer, how to reclaim the slopes of a 50 acre gravel pit.

The gravel pit is a great open bowl, east facing, with slopes rising as high as 80 feet. The design provides three private streets running parallel with the slopes in the form of loop roads connecting to the main avenue. The streets provide access to the housing at the top and bottom ends of each structure.

The structures, all earth-sheltered, are either facing southwest or northeast to allow optimum use of all of the slopes on the site. Only one face of each structure is exposed as a conventional facade, and each unit has a private yard space and is defined as private property in an attempt to offer the amenities of single family detached housing which characterizes the adjacent communities.

An interesting aspect of this project involves the owner's business as a concrete block manufacturer. He can act as his own developer, take gravel from the site, make concrete blocks and return it to the site as the major material of the project — a modern day parallel to adobe construction.

In addition to earth-sheltered construction, "on-site/in-house" production of the major building material is another means of conserving energy.

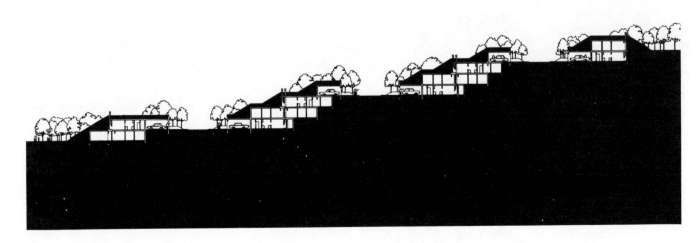

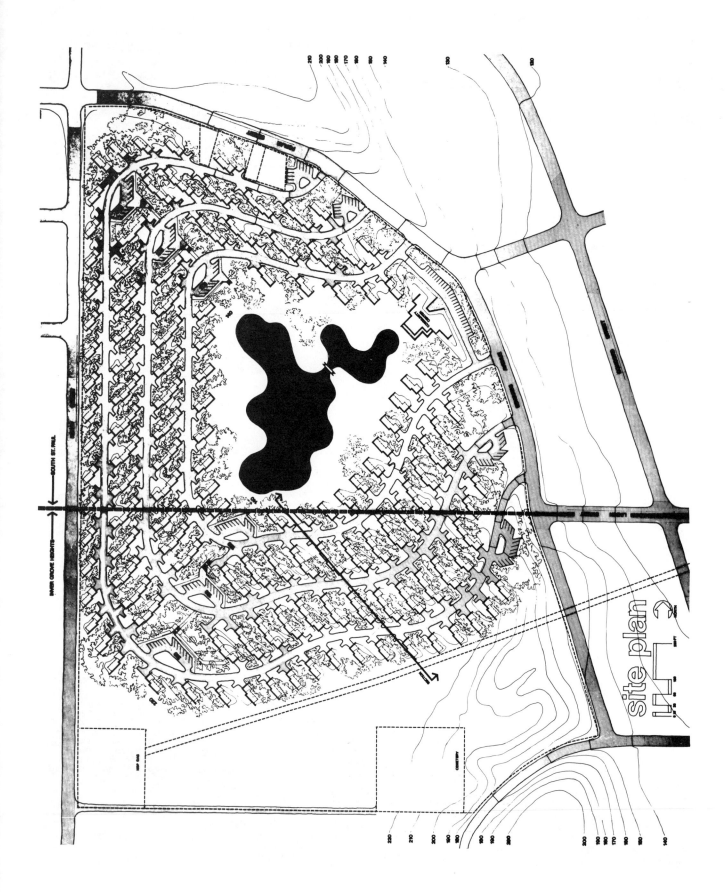

site plan

upper level

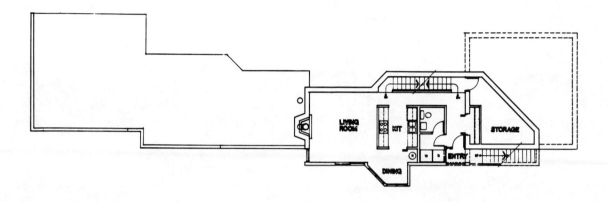

middle level

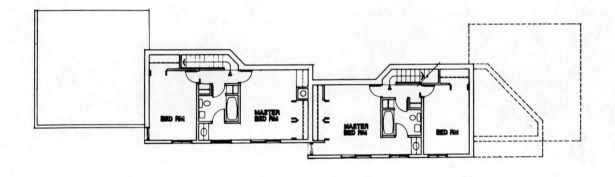

lower level

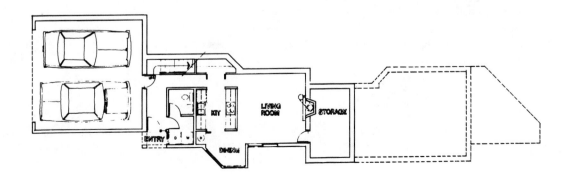

95

# Second Award

## Condominium Complex for an Oklahoma Site
by Allen L. Brown, Student, Oklahoma State University, Stillwater, Oklahoma

The program for this project specified a condominium complex which would: create a sense of community but retain privacy; provide recreation and entertainment opportunities; provide protection from violent storms; appeal to consumers; and minimize dependence on artificial environmental control systems. Energy conserving features focusing on the latter goal include site design, materials chosen for thermal storage capacity, space zoning and orientation designed to control solar effects and the use of the earth as a "heat sink."

The entrance road to the complex is along the north edge of the tree cover. On-site trees will be saved where possible and used to shade the road. The units will face this road and are built into the south-facing slope in groups. The interlocking and staggering of the units produces high densities while retaining privacy, individual entrances, views, access to sunlight and direct contact with the earth. The design also creates a sense of community by "weaving" the units into a neighborhood. The trees and man-made ponds to the south of the units provide natural cooling as the prevailing south winds take advantage of plant transpiration and the water evaporative effect to lower the air temperature. Northwest winter winds are deflected up and over the complex by the adjacent north slope and should have little impact on heat loss. The tree cover and pond area also act as a recreational area to include nature and jogging trails, fishing, and a future clubhouse. Extra parking for large parties will be above the complex accessed at the west end of the site and connected by steps to the entrances.

The building design is based on two spatial concepts. First, units are zoned into morning and afternoon use areas, with upper level morning sleeping areas directed toward the southeast light while the lower level afternoon living areas face the southwest light. Second, the center of the unit is considered to be the dining "hearth." This is expressed by unifying the morning and afternoon

areas through this central, skylighted, two-story space. The entrance and parking are at the lower level of the southwest side. The individual units have outdoor terrace areas for private use. The earth-sheltered structure, overhangs, and setbacks protect the occupants from storm damage.

The overhangs on the southwest and southeast are designed to provide complete shading of glass areas during critical midday hours. The staggered configuration of the units provides early morning and late afternoon shade for adjacent units when the summer sun rises and sets about 30° north of east-west. The skylight shutters direct south sunlight by automatically tilting at an angle to intercept the sun's rays while permitting indirect light to enter. Heat generated by solar radiation trapped in the skylight, along with that rising from below, will be stored in the brick "eggcrate" and recovered for domestic hot water use. Floor and lower wall area in direct contact with the earth receive "free cooling" due to the effect of air and ground temperatures. The earth contact is maximized by offsetting the floor areas, minimizing party walls between units, and using a folded plate two-story retaining wall as a functional and symbolic "heat sink." The unit will be ventilated at night to exhaust hot air and closed during the day to take advantage of cooler air brought in at night. Air convection will be generated as warm air is cooled along the rear wall and falls to the lower spaces, preventing air stagnation. Estimated heat gain in summer is 22,336 BTUH.

The overhangs also are designed to permit the low angle of the winter sun, which rises and sets at about 30° south of east-west, to enter the space and strike the floor. This radiation is absorbed by either a brick or concrete masonry floor and radiated into the space at night. Fireplaces are equipped with "heatilator" type devices and contribute heat at both the upper and lower level. The moderate temperature of the earth also radiates heat into the space. Estimated heat loss in winter is 18,442 BTUH.

Oklahoma's dust storms make natural ventilation a questionable alternative to mechanical forced-air systems at most times in the year. However, the prevailing summer winds will be cooled as mentioned above and can be utilized during the periods of temperate outdoor conditions. This cooled breeze can be channeled through the southwest and southeast wall openings and drawn out through the upper level operable louvers, augmented by negative pressure created as the wind passes over the cavity between the units. Northwest winter winds are deflected over

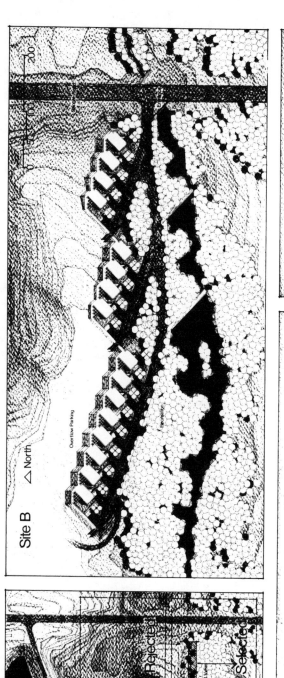

Site B △ North

Overflow Parking

Recreation

Site Analysis

Site A

Rejected

Selected

Site B

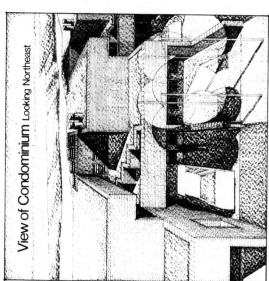

View of Condominium Looking Northeast

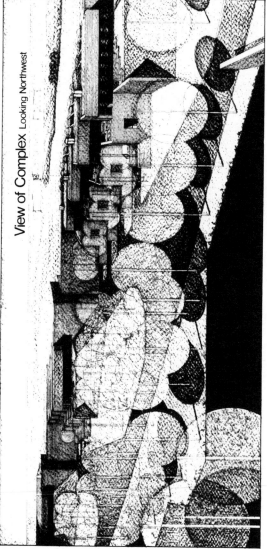

View of Complex Looking Northwest

the top of the units by the north slope and have no direct contact with the walls of the units, but also create negative pressure drawing air through the louvers for exhaust.

Daylighting is accomplished through the use of skylights over the central dining area, considered to be in use the entire day. The skylight shutters control the amount and direction of direct and indirect sunlight. This results in seasonal variations in reflected light washing down the back wall. The morning and afternoon areas receive daylight directly when the use of the space is greatest. The depth of spaces along exterior glass walls is 16 feet, less than the common-practice of two and one-half times the window height of eight feet. The southeast lower-level wall also serves to reflect diffuse light into the kitchen in the morning. This combination of perimeter wall openings, roof openings at "buried" spaces and reflected light results in uniform natural daylighting and little need for artificial lighting during the day.

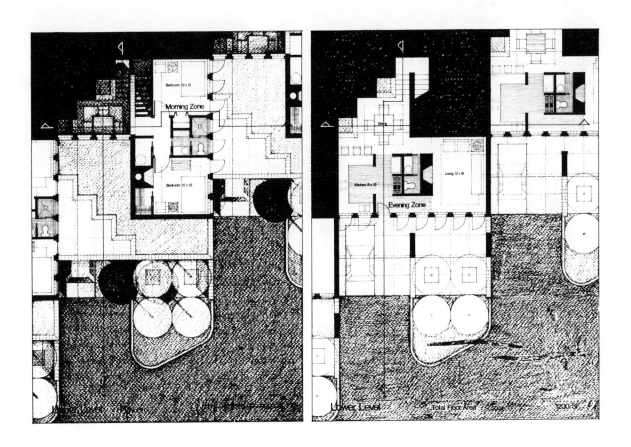

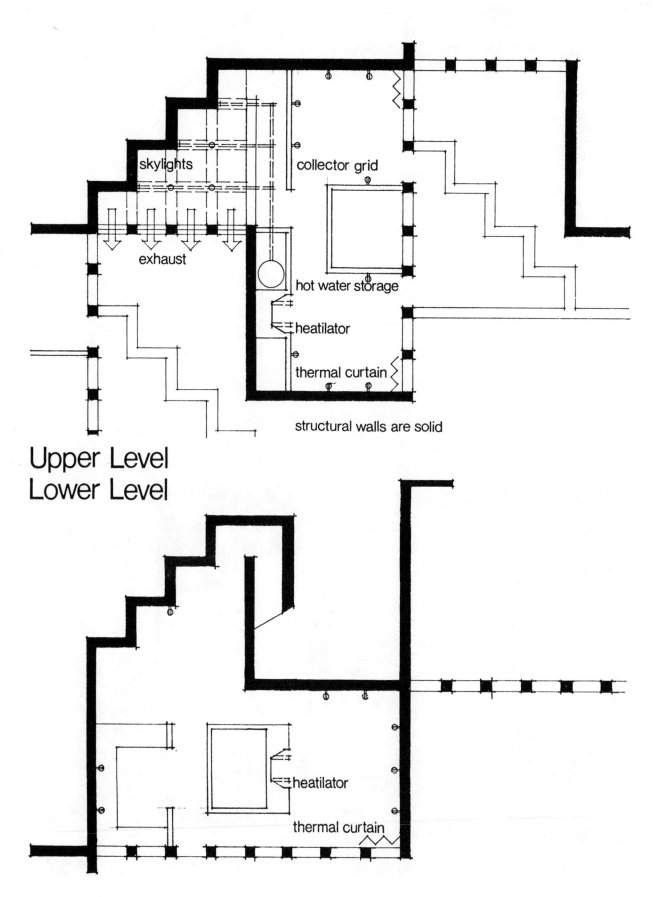

skylights

collector grid

exhaust

hot water storage

heatilator

thermal curtain

structural walls are solid

## Upper Level
## Lower Level

heatilator

thermal curtain

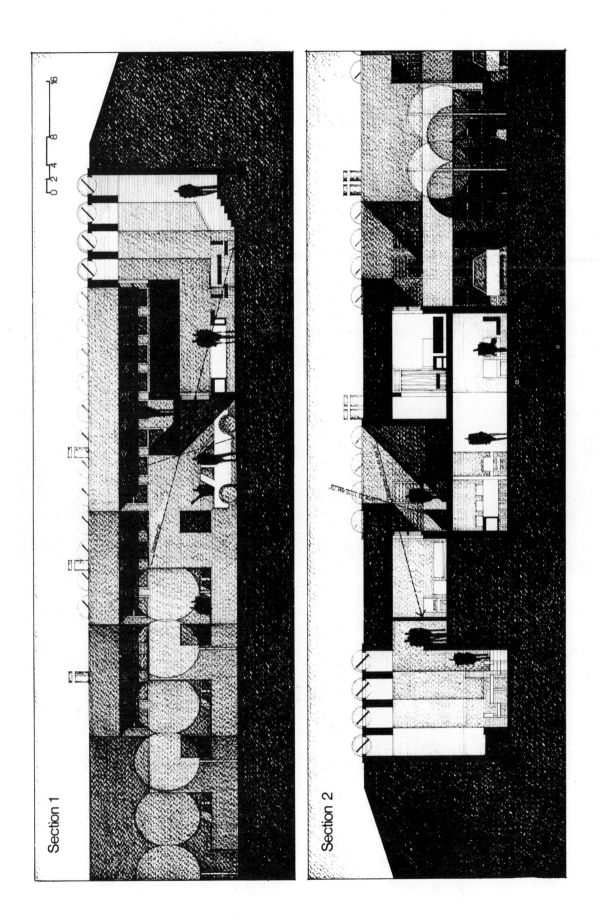

Section 1

Section 2

0 2 4    8    16

# Merit Award

## Low Cost Complex on a Highly Visible Site in Aspen, Colorado
by David Gibson, Pielstick Gibson Reno, Aspen, Colorado

The program required the development of 24 "low-income" employee units on a high-rent, high-visibility section of Main Street in a fashionable mountain resort.

A difficult political situation was resolved through the use of an earth-sheltered solution. The welfare of "employees" is a popular cause until it threatens someone's property values.

However, this earth-sheltered solution gives something to everyone: to the employees it offers a pleasant place to live with small utility bills. To the immediate neighbors it provides an apparent reduction in density and an increase in highly-prized open space. To the public at large, hooray — it seems to be another park. To the client, it affords an opportunity to establish a reputation as an innovator in the community.

The poured-in-place concrete walls and pre-cast decking provide a stable base for rigid insulation, waterproofing and a 250 pound live load of sod and berms. Above grade, brick is used as a finish material. The primary finish material, however, is carpet bugle, white dutch clover, rye grass/Kentucky bluegrass, juniper, aspen and blue spruce, with built-in sprinklering to counteract the arid climate.

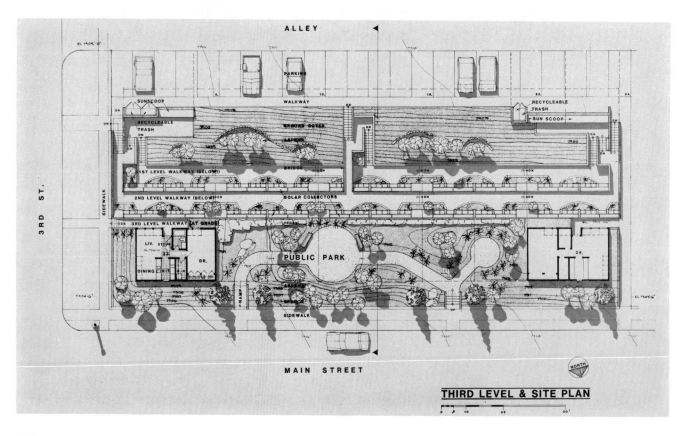

THIRD LEVEL & SITE PLAN

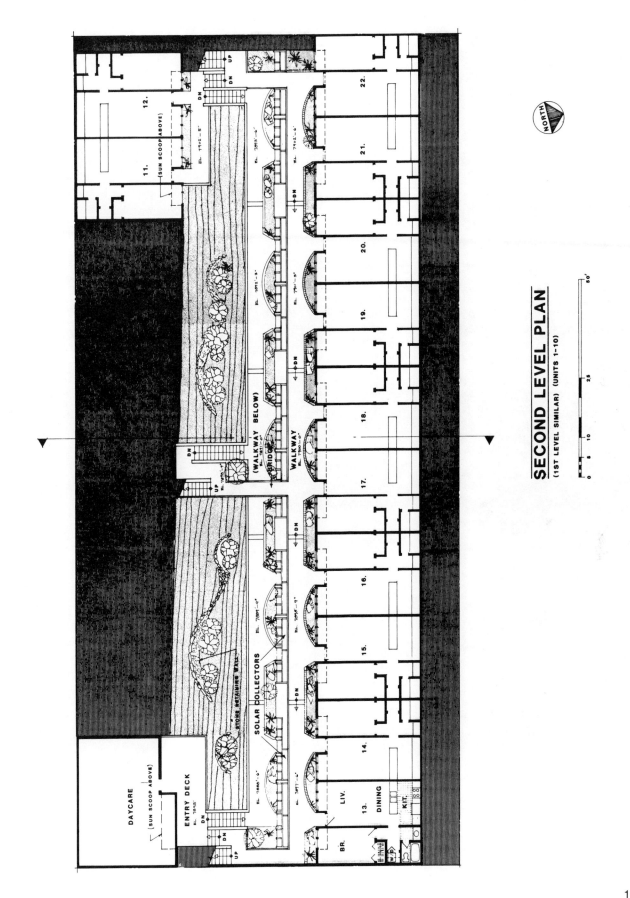

**SECOND LEVEL PLAN**
(1ST LEVEL SIMILAR) (UNITS 1-10)

NORTH

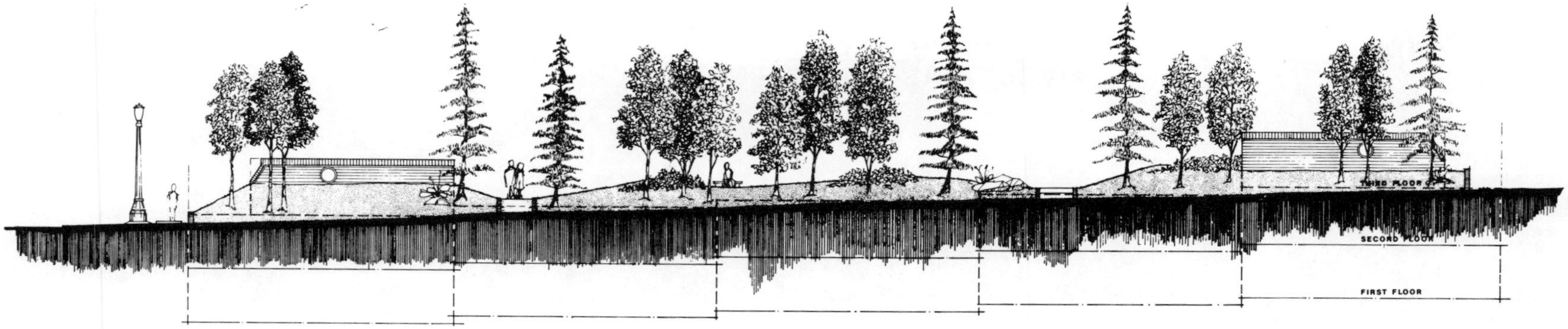

NORTH (MAIN STREET) ELEVATION

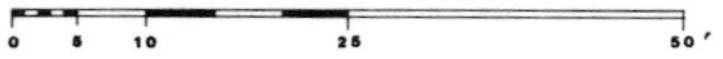

0 5 10 25 50'

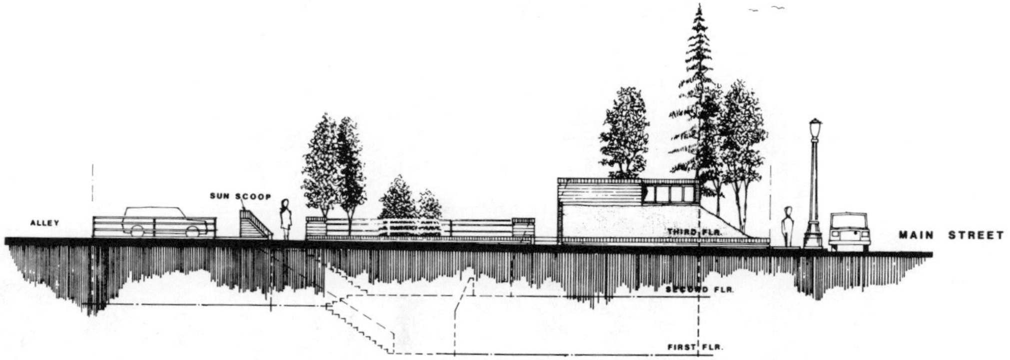

EAST (THIRD STREET) ELEVATION

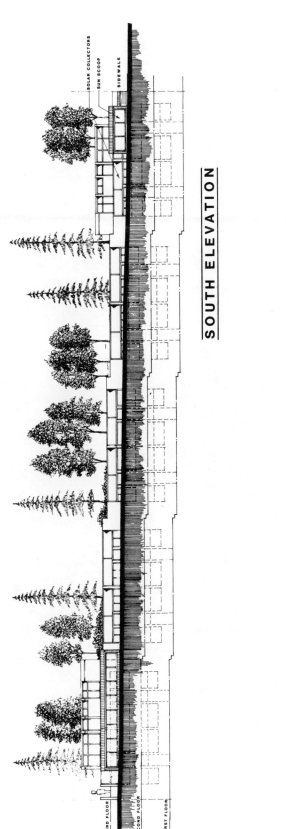

SOUTH ELEVATION

SOLAR COLLECTORS
SUN SCOOP
SIDEWALK

THIRD FLOOR
SECOND FLOOR
FIRST FLOOR

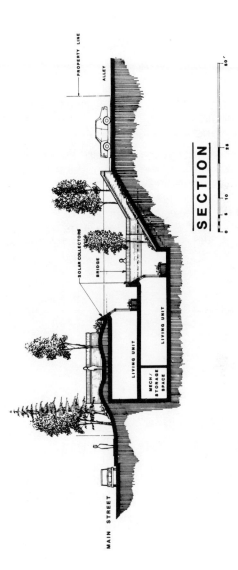

SECTION

PROPERTY LINE
ALLEY

SOLAR COLLECTORS
BRIDGE

LIVING UNIT
LIVING UNIT
MECH/
STORAGE
SPACE

MAIN STREET

0  5  10  25  50'

105

# Honorable Mention

## Row House Complex for Wilmington, Ohio by Andrea Fossati, Student, Ohio State University, Columbus, Ohio

These attached single family houses are part of a project of 40 passive solar residences and a passive solar commercial center at Wilmington, Ohio. The site is surrounded by a forest to the north, existing traditional single-family houses on the north, south and east, a street to the south and open space and a lake to the southwest.

Twenty-two earth-sheltered residences are located near the lake, 18 above-ground residences are located to the north and an earth-sheltered commercial building is to be constructed southeast of the lake, adjacent a local street. Locating the earth-sheltered buildings near the shore allows all residences a view of the lake. In addition, the use of stone and wood for the exteriors makes these new architectural forms compatible with existing houses in the area.

Utilization of the hill for shelter and earth-covered roofs decreases the heating requirements approximately 45%. The sun provides 60-70% of the heating needs of the complex. Specifically, the greenhouse and balconies collect solar energy which is absorbed by the concrete walls and tile flooring.

This passive solar system offers multiple advantages. The greenhouse and the balconies are extensions of the dwelling space as well as efficient parts of the heating system. The simplicity of the system means long-life and freedom from maintenance.

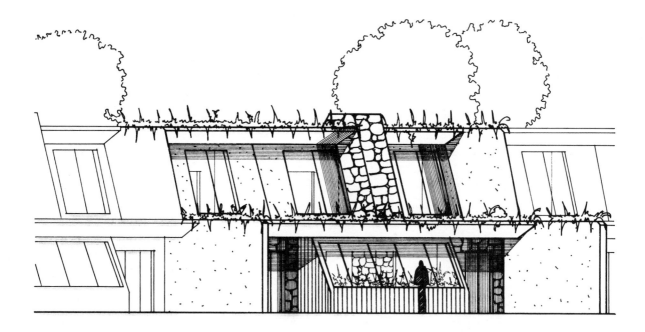

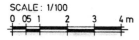

SCALE : 1/100

0  05 1    2    3    4 m

# SITE PLAN

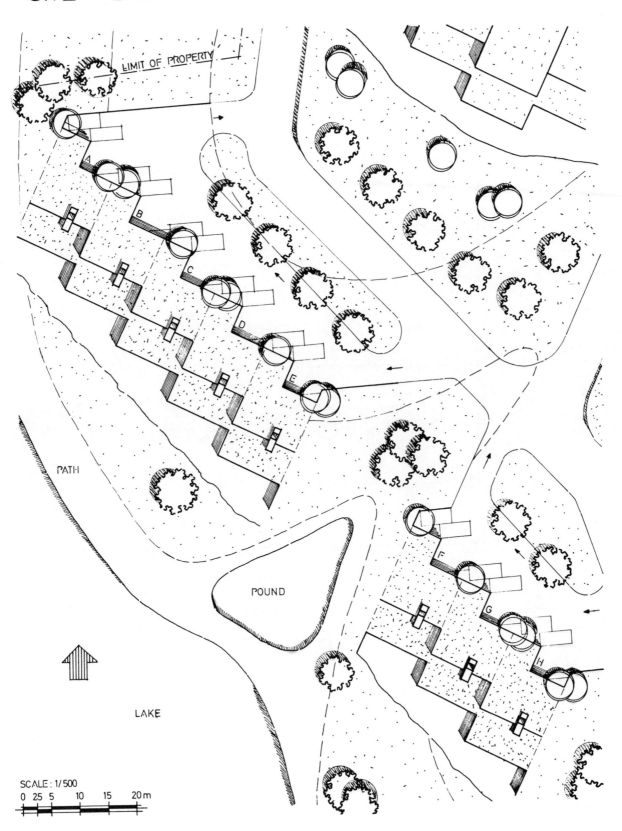

LIMIT OF PROPERTY

A

B

C

D

E

F

G

H

PATH

POUND

LAKE

SCALE : 1/500

0  2·5  5      10      15      20 m

# FIRST LEVEL / ENTRY

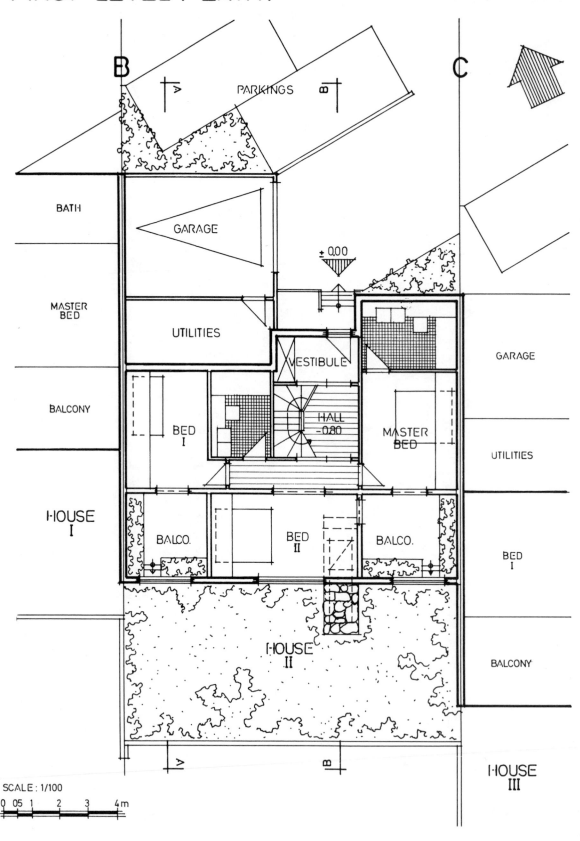

B

PARKINGS

C

BATH

GARAGE

± 0.00

MASTER BED

UTILITIES

VESTIBULE

GARAGE

BALCONY

BED I

HALL −0.80

MASTER BED

UTILITIES

HOUSE I

BALCO.

BED II

BALCO.

BED I

HOUSE II

BALCONY

HOUSE III

SCALE : 1/100

0  05  1      2      3      4m

# SECOND LEVEL / GARDEN

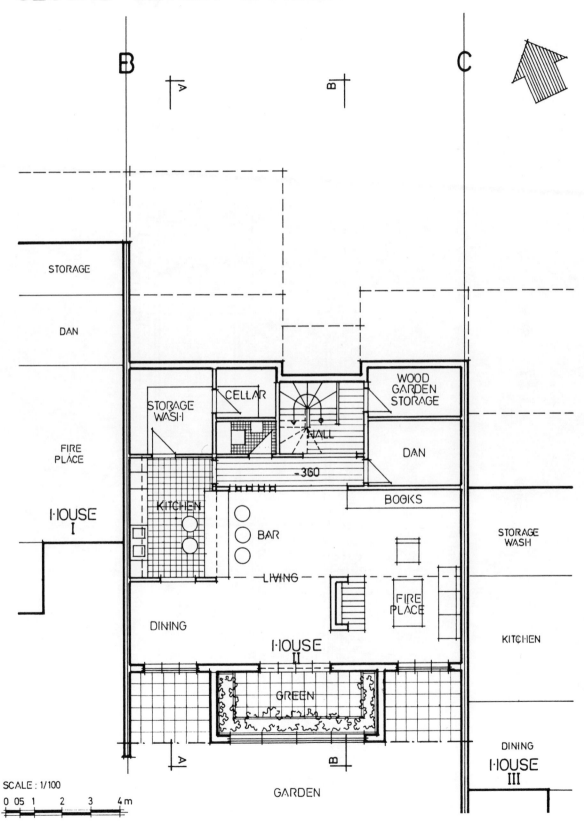

B

A

B

C

STORAGE

DAN

FIRE
PLACE

HOUSE
I

STORAGE
WASH

CELLAR

WOOD
GARDEN
STORAGE

HALL

DAN

-360

BOOKS

KITCHEN

BAR

STORAGE
WASH

LIVING

FIRE
PLACE

KITCHEN

DINING

HOUSE
II

GREEN

A

B

DINING

HOUSE
III

GARDEN

SCALE : 1/100

0  05  1    2    3    4 m

# SECTION A-A

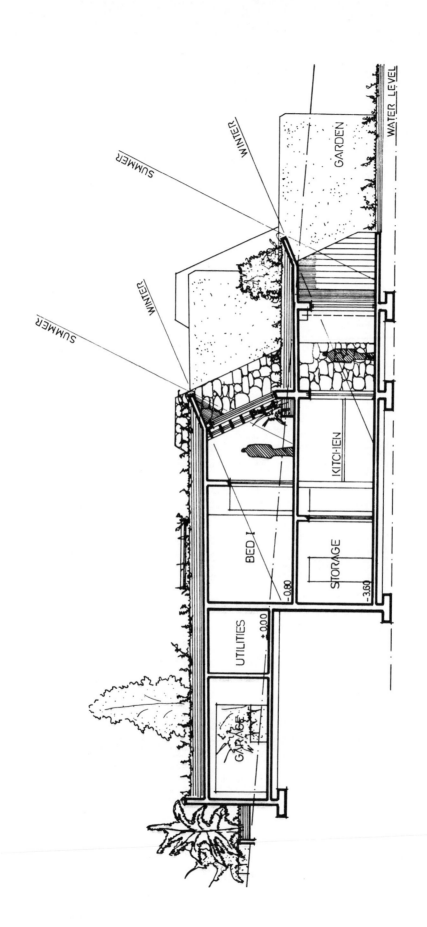

WINTER

SUMMER

WINTER

SUMMER

GARDEN

WATER LEVEL

KITCHEN

BED I

STORAGE

-0.80

-3.60

UTILITIES

+0.00

GARAGE

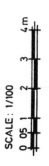

SCALE: 1/100

0 05 1   2   3   4m

# SECTION B-B

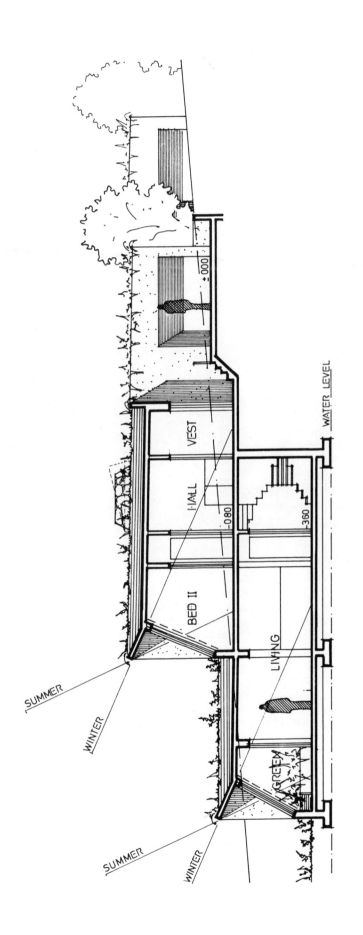

VEST

HALL

BED II

SUMMER

WINTER

+ 000

- 0.80

- 360

LIVING

GREEN

WATER LEVEL

SUMMER

WINTER

SCALE : 1/100

0 0.5 1   2   3   4 m

## A Prototype Underground Housing Kit for North America
by Michael O. Winters, Designer, Lakewood, Colorado

This program offers a theoretical prototype for an energy responsive living unit that could be built, with suitable variations, over much of North America. The concept is for a unit that would be sold like a condominium, at a price that would be among the lowest in a given market, and that would have minimum operating costs.
A site, typical of many to be found across America, was selected and used to develop solar design principles for two opposite climate zones — a cool climate, representative of Madison, Wisconsin, and a hot arid climate, representative of Phoenix, Arizona.

One design solution can be adapted to all climate types with only slight variations to the type of energy system, overhang, landscaping, etc. due to the moderating effect on heating and cooling temperature differentials of underground structures.

A passive/hybrid heating and cooling system was developed and can be adapted to any climate, simply by choosing the appropriate "south facade" kit. The south facade acts as a "sealed" solar collector, providing heat for the building module. An electric baseboard-fan unit located within the south facade acts as a back-up heat source. The fan unit provides additional pressure for air circulation in the plenum spaces of the facade and precast floor and roof planks.

Underground air ducts which pass warm outside air through cooler subsurface earth temperatures provide necessary cooling to the building modules in three ways: cool air can be supplied directly into the living space as it is needed; cool air can be flushed through the plenum in the concrete roof planks; lowering the mean radiant temperature of the mass; a combination of cool air supply and radiant cooling can be used to cool the living spaces.

Thus, the building becomes a mechanical plenum, heating or cooling the interior spaces through storage and radiation of the concrete thermal mass.

The following heating efficiencies were calculated for the two opposite climate types: Madison, Wisconsin — 85%; Phoenix, Arizona — 100%.

The ideal building system suited to this underground solar prototype is based on precast concrete components in the form of a building "KIT" which uses a simple module.

In addition to the energy benefits gained by placing the building modules underground, specific site and architectural benefits were also realized. Zoning and height restrictions can be a problem when dealing with a small typical site. By building underground, height problems associated with the density of this project are eliminated and the open space and views of the site are totally preserved and shared equally by each unit. Rooftop terraces, developed at grade level allow maximum sun exposure and provide exterior private space for each unit. Alley level parking is hidden from the public street view through the use of a building storage element. This element acts as a doorway to the units below, separating public space from private space. The motif of this element is repeated at each level, defining entry, circulation and the private area of each unit, as well as providing structural support for the addition of (optional) domestic hot water solar collectors.

studio

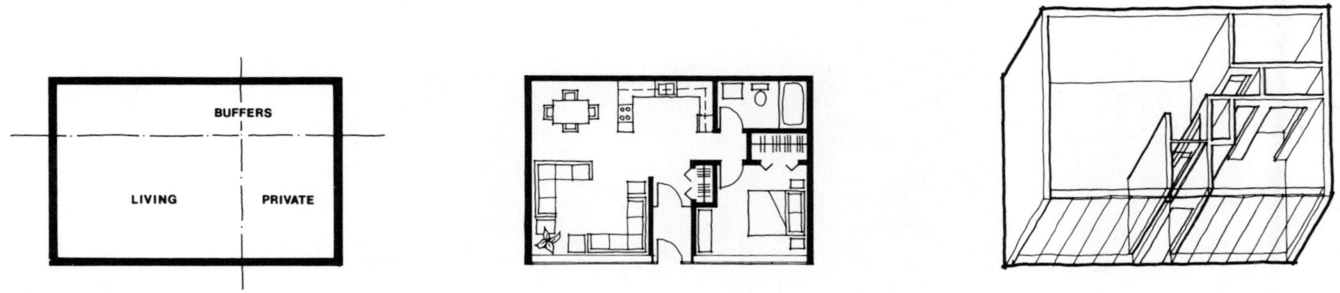

1 br.

2 br.

living units

BUFFERS LIVING PRIVATE

BUFFERS LIVING PRIVATE

BUFFERS LIVING PRIVATE

0 8 16

113

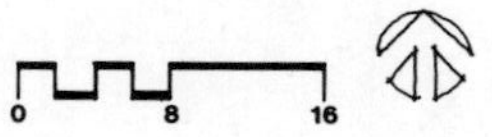

# the module kit

## 20 x 20 module

- THERMAL MASS ——— 530 c.f.
- SOUTH APERTURE ——— 140 s.f.
- VOLUME ——— 2800 c.f.

OVERHANG-
ADAPTS TO CLIMATES.

- 4 ft. PRECAST ROOF PLANKS.
- PRECAST BEAM
- 8 in. PRECAST BEARING WALLS.
- 4 ft. PRECAST FLOOR PLANKS.

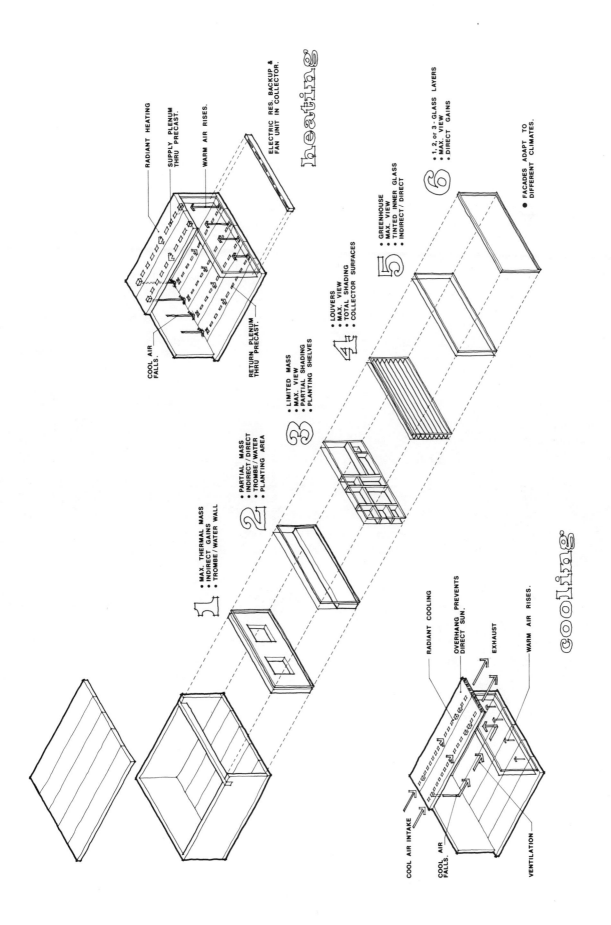

**heating**

RADIANT HEATING

SUPPLY PLENUM
THRU PRECAST.

WARM AIR RISES.

ELECTRIC RES. BACKUP &
FAN UNIT IN COLLECTOR.

COOL AIR
FALLS.

RETURN PLENUM
THRU PRECAST.

**6**
- 1, 2, or 3 - GLASS LAYERS
- MAX. VIEW
- DIRECT GAINS

- FACADES ADAPT TO
DIFFERENT CLIMATES.

**5**
- GREENHOUSE
- MAX. VIEW
- TINTED INNER GLASS
- INDIRECT / DIRECT

**4**
- LOUVERS
- MAX. VIEW
- TOTAL SHADING
- COLLECTOR SURFACES

**3**
- LIMITED MASS
- MAX. VIEW
- PARTIAL SHADING
- PLANTING SHELVES

**2**
- PARTIAL MASS
- INDIRECT / DIRECT
- TROMBE / WATER
- PLANTING AREA

**1**
- MAX. THERMAL MASS
- INDIRECT GAINS
- TROMBE / WATER WALL

**cooling**

RADIANT COOLING

OVERHANG PREVENTS
DIRECT SUN.

EXHAUST

WARM AIR RISES.

COOL AIR INTAKE

COOL AIR
FALLS.

VENTILATION

**adaptable south facade kits**

115

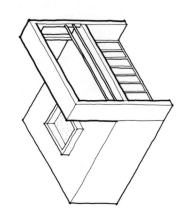

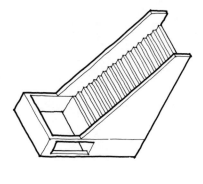

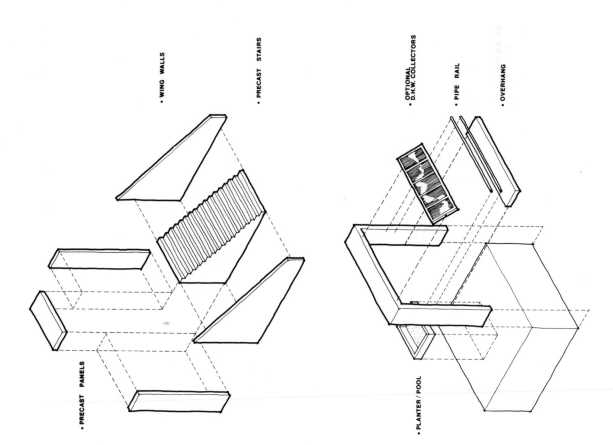

accessory kits

- WING WALLS

- PRECAST STAIRS

- OPTIONAL D.H.W. COLLECTORS

- PIPE RAIL

- OVERHANG

- PRECAST PANELS

- PLANTER / POOL

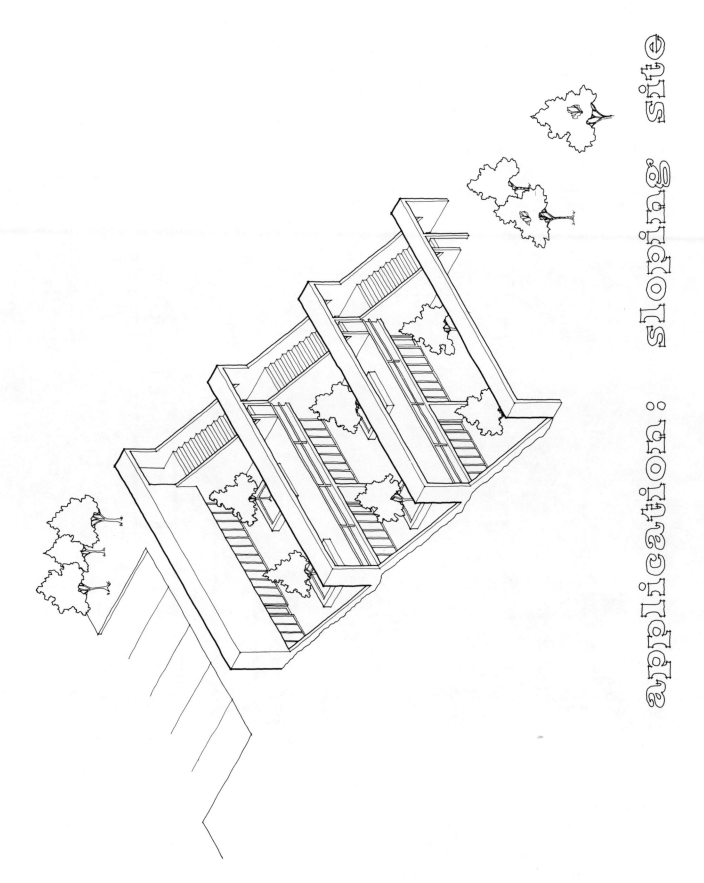

application: sloping site

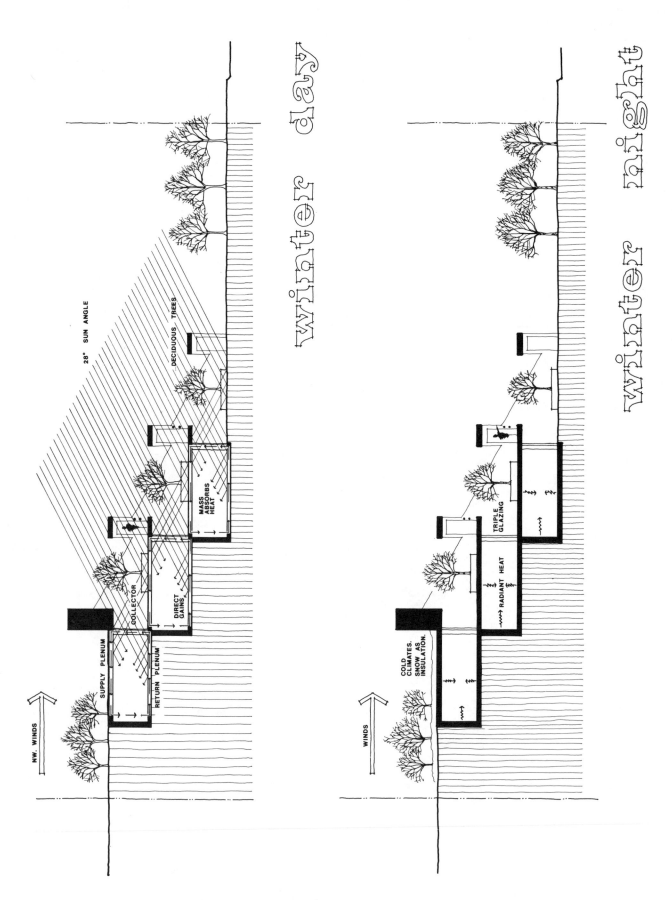

winter day

28° SUN ANGLE

DECIDUOUS TREES

MASS ABSORBS HEAT

COLLECTOR

DIRECT GAINS

SUPPLY PLENUM

RETURN PLENUM

NW. WINDS

winter night

TRIPLE GLAZING

RADIANT HEAT

COLD CLIMATES. SNOW AS INSULATION.

WINDS

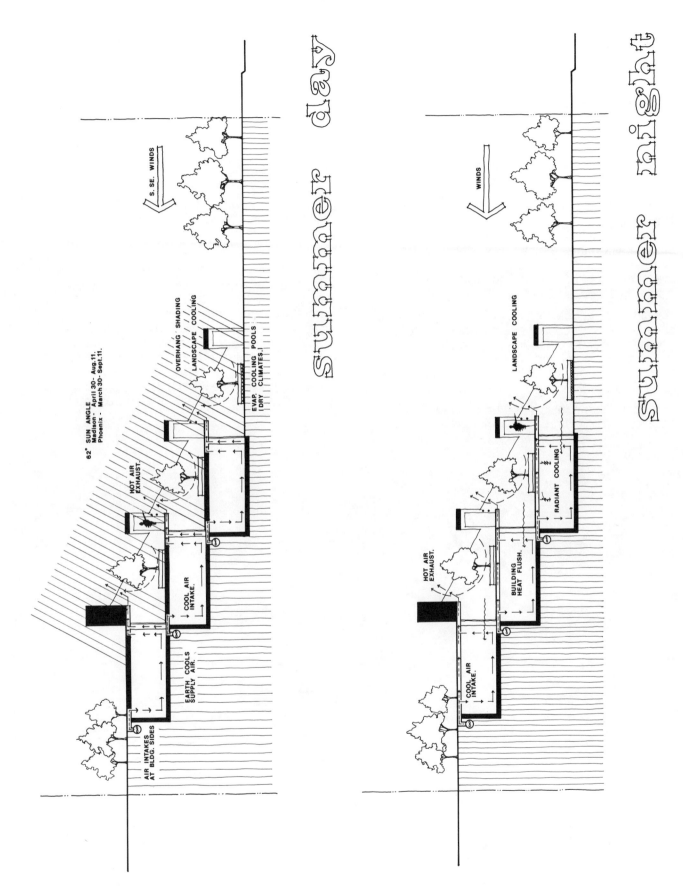

summer day

62° SUN ANGLE
Madison - April 30 - Aug. 11.
Phoenix - March 30 - Sept. 11.

OVERHANG SHADING
LANDSCAPE COOLING

EVAP. COOLING POOLS
(DRY CLIMATES.)

HOT AIR EXHAUST.

COOL AIR INTAKE.

EARTH COOLS SUPPLY AIR.

AIR INTAKES AT BLDG. SIDES

summer night

LANDSCAPE COOLING

WINDS

RADIANT COOLING

HOT AIR EXHAUST.

BUILDING HEAT FLUSH.

COOL AIR INTAKE.

# Honorable Mention

## Prefabricated Kibbutz Complex for the Arava Area of Israel
### by Rafael Danon, Student, New York Institute of Technology, New York, New York

The Arava in southern Israel is a relatively unsettled desert area. This valley is bounded by mountains on the east and west which separate Israel and Jordan. The Arava makes contact with the Dead Sea to the north (the lowest level in the world) and the Red Sea to the south.

The unique attributes of the area — the hot climate, the high cost of energy, nearness to a potentially dangerous border — are factors which led to a solution of earth-sheltered housing. Remoteness from the cities and the high cost of labor make prefabrication an asset.

The basic plan consists of four duplexes connected by a fifth common duplex in the form of hexagons. This plan provides for two small apartments on each level or larger duplex apartments. However, the flexibility of precast design permits six variations in apartment size (i.e. ranging from 425 square feet to 1040 square feet) to accommodate one person to families with four or five children.

The building consists of a complex of four basic structures which can be utilized as either eight large apartments, 16 small apartments or a combination of the two. Each apartment has a minimum of one bedroom, bathroom, dining space, kitchen, and balcony. Within each complex is a common play area.

The first level will be constructed approximately six feet below the ground. The earth removed from the foundation will be used to cover exposed parts of the building. The roof, which should be covered by two feet of earth, will be used for gardening and will have an underground, automatic system for irrigation which also provides additional cooling for the house. More extensive cooling is achieved through cross ventilation plus ceiling fans on top floors and individual vapor coolers in each apartment.

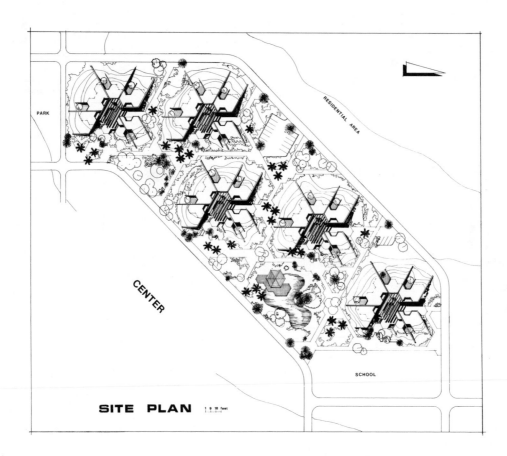

PARK

RESIDENTIAL AREA

CENTER

SCHOOL

**SITE PLAN** 1 9 18 feet

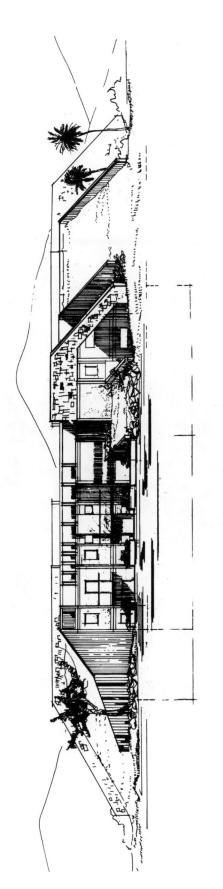

**EAST ELEVATION**

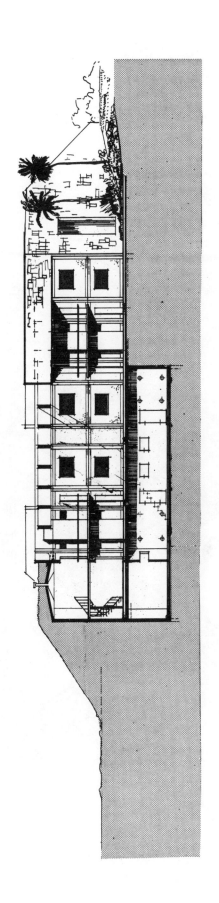

**SECTION**

1 4 8 feet

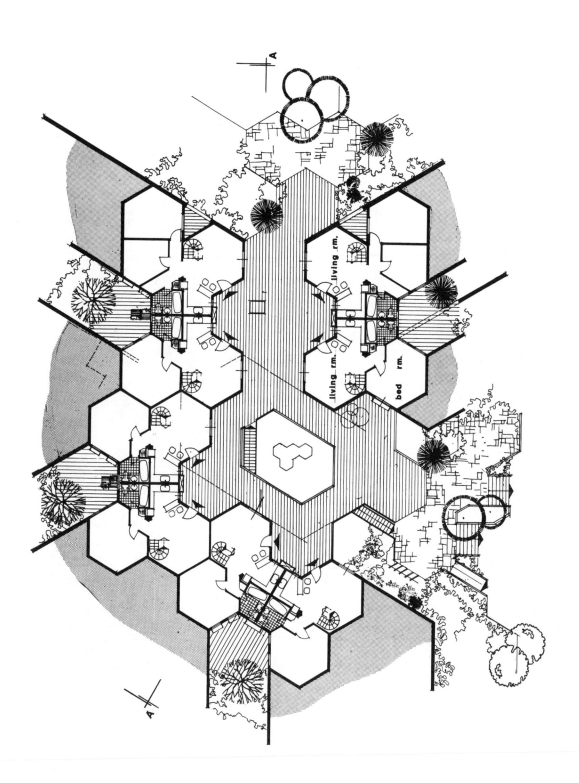

living rm.

living rm.

bed rm.

**FIRST FLOOR** 1 4 8feet

122

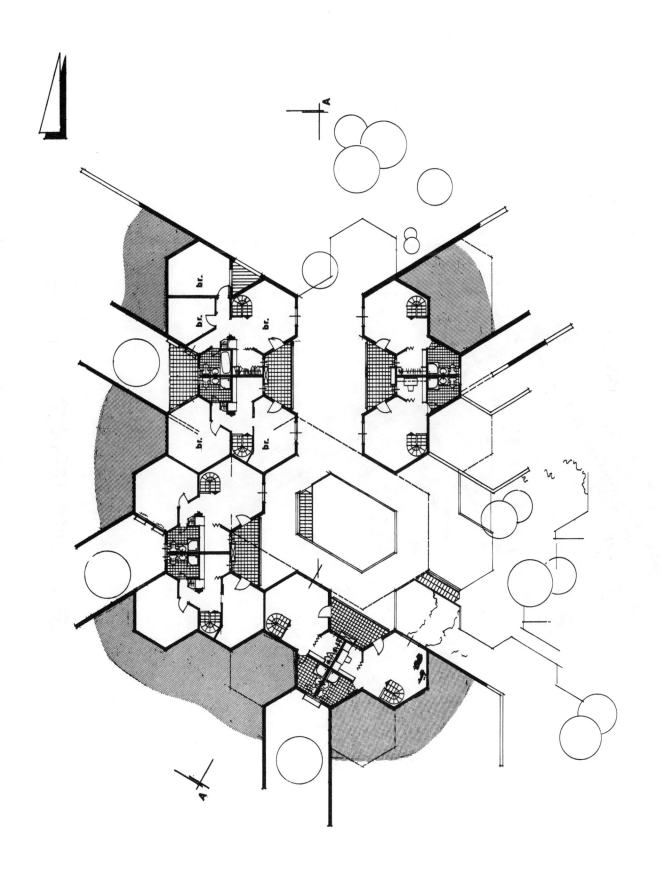

**SECOND FLOOR** 1 4 8 feet

br. br. br. br. br. br.

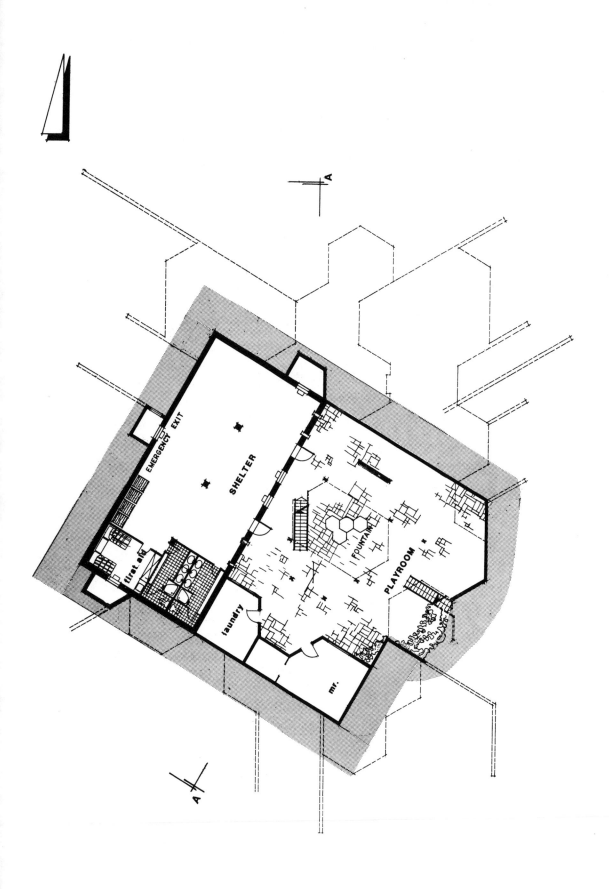

**BASEMENT**

1 4 8 feet

125

# Earth-Sheltered Mobile Home Park Prototype
## by Robert Vilkas, Principal, SPACEMAN/ architecture and planning, Dayton, Ohio

Mobile homes, by necessity, are here to stay. Their low cost and mobility make them a natural choice for a great many Americans. However, the mobile home and the planning of the traditional mobile home "park" have made serious compromises on the development of a quality living environment.

Usually, these units are crowded so closely together that privacy is non-existent. Fire can spread quickly from unit to unit and the minimal construction standards make these units especially susceptible to damage from wind and tornadoes. Finally, despite their efforts, the designers of these manufactured homes cannot disguise the truth. A trailer is a metal box on wheels and the less you see of it, the better.

The purpose of this hypothetical proposal is to address these undesirable attributes and develop a scheme that not only solves the technical problems but also affords a more desirable living environment.

A Mobile Underground Community would consist of earth-sheltered lots sized to accommodate industry-standard units. The plans presented herein show all single-wide units, but double wides could be accommodated in lots sized appropriately. The actual units would only need slight modifications from those currently available. First, entry and exit must be at the front and rear walls only. Second, use of skylights, detachable clerestories, or lightwells would be provided to introduce natural light within interior rooms.

Units would be positioned between ten foot high concrete walls and an adjustable, insulated apron would close off the crawl space as well as the space between the unit and the concrete walls, thus creating a 12 inch air space wrapping the unit which would reduce heat loss from the unit significantly.

The additional advantages of this approach are numerous. The earth-berming between the units would provide privacy, since both front and rear courts would be screened from adjoining lots. The spread of fire between units would be nearly impossible and the unit's resistance to severe weather damage greatly increased.

Finally, the aesthetics of such a community would be several levels above that normally found. The front and rear courts could be landscaped to individual tastes, while the earth berms could be planted with a ground cover that would minimize maintenance, introduce natural colors between the units, and visually tie the whole development together.

Obviously, site development costs for such a project would be considerably higher than traditional mobile home parks, but the improved quality of living environment, the attractiveness of energy efficiency, and the possibility of locating such a development closer to established residential areas could justify increased rental rates, thus offsetting initial investments.

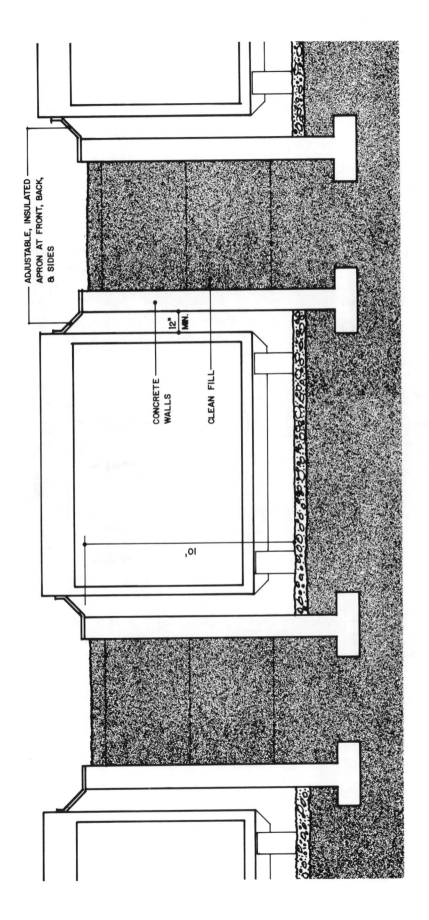

ADJUSTABLE, INSULATED APRON AT FRONT, BACK, & SIDES

CONCRETE WALLS

CLEAN FILL

12" MIN.

10'

0  1        3        5

TYPICAL CROSS·SECTION

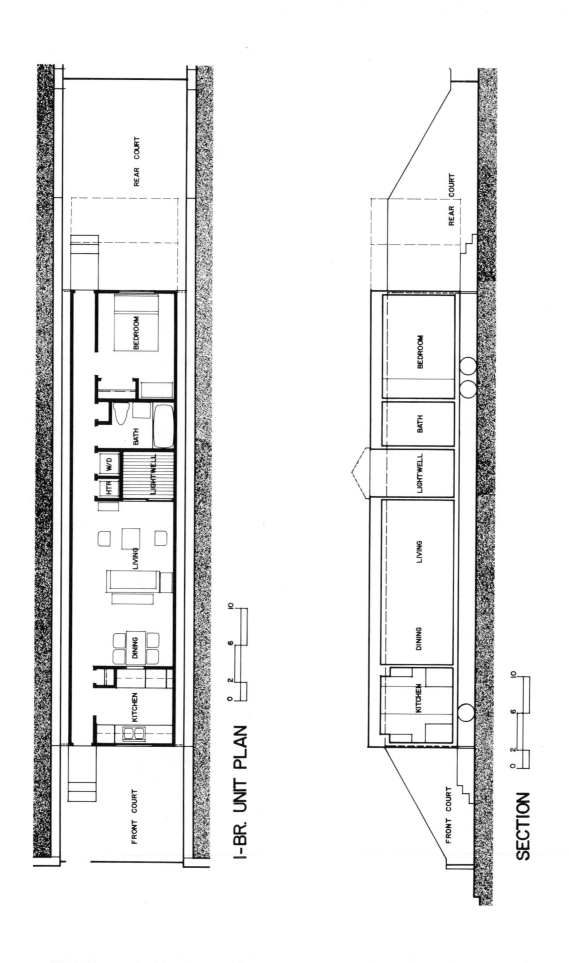

I-BR. UNIT PLAN

FRONT COURT

KITCHEN

DINING

LIVING

HTR
W/D

LIGHTWELL

BATH

BEDROOM

REAR COURT

0  2    6    10

SECTION

FRONT COURT

KITCHEN

DINING

LIVING

LIGHTWELL

BATH

BEDROOM

REAR COURT

0  2    6    10

CUL·DE·SAC   ARRANGEMENT

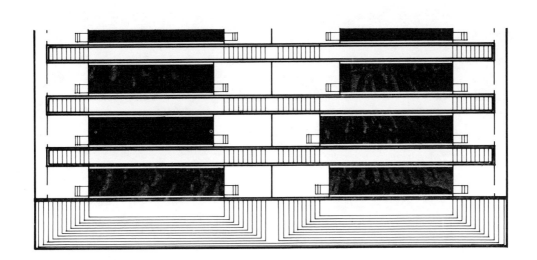

ROW·HOUSE   ARRANGEMENT

# Commercial

"There was more variety in the Commercial work by both students and professionals. However, the greater experience of the professionals came through strongly in this category. Surprisingly, professional work on the whole was far more exciting than student work. One usually thinks of professionals as being tied down by having to face the possibility that their work may be built. In spite of this, the professionals were actually the more adventurous."

Edward Allen

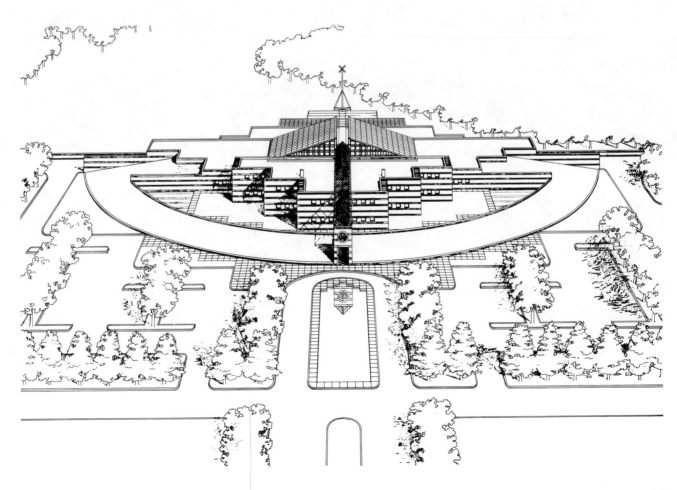

# Merit Award

## Visitor/Interpretive Center for a California State Wild Flower Reserve
by S. Pearl Freeman and Robert D. Colyer, The Colyer/Freeman Group, San Francisco, California

After eight years of acquiring 1,671 acres of desert wildflowers in the high western Mojave Desert (elevation 3000 feet), the client agency from the State of California and residents of the area initiated the program for the design of the Jane S. Pinheiro Visitor/Interpretive Center for their new State Park and Wildflower Reserve. The building is intended to house permanent and temporary exhibits of desert wildflowers and ecology, including the wildflower paintings of a prominent local artist. It is to be an inviting public building, one that is not an institutional distraction in the natural environment. In addition, the center is to serve as a demonstration project of "state-of-the-art" appropriate technology systems within a modest fixed budget.

The architect's approach was initially inspired by the indigenous desert construction of the American Southwest. The building is designed to blend both physically and historically into the rolling buttes while responding to the variable climatic conditions of the high desert. Matching the desert soil color, horizontal bands of smooth block pierce the predominantly split-face block finish suggesting the layers of native sedimentary rock. Exterior wing walls step and curve where the building meets the earth. This earth-sheltered characteristic is especially crucial to energy efficiency, acting as a constant moderator at ±56°F against the summer highs of 113° and winter lows of 3°.

The exhibit room incorporates both natural and artificial track lighting to accommodate the changing and varied displays. A 30 foot north clerestory along the rear wall allows daylight to be the primary source of light, without the introduction of significant heat gain. The symmetrical niches flanking the daylit rear wall frame two of the permanent displays in the center — the video exhibit and the evaporative cooling fountain. Daylighting is also the major source of light in the office/vestibule area. Backed by a skylight and floor-to-ceiling windows, the stained glass windows at the sales counter will be a dramatic focus as one enters the building. The stepped/eroded wall in the same area visually unites the two spaces while opening for the passage of ventilation air and daylight from the exhibit room. With the expectation of high public use and additional funds, the 1900 square foot building was designed to be expanded from either the east or west without violating the original design concepts.

The intent of a demonstration project demands natural energy systems that are both passive and observable. Like its desert predecessors, the visitor center will use the sun's energy, the vigorous winds and the stable temperatures of the earth to provide its heating, electricity and cooling needs. 100% of the building's heat will come from direct-gain/mass-storage in the exhibit/office/vestibule areas and a trombe wall in the restrooms. Computer analysis shows that during the winter months, using R-5 night insulation at the glazing areas, the building will maintain 67° during the visitor center use hours. Summer cooling of the exhibit areas will be achieved by night-time pre-cooling of the mass using a 150 foot long underground duct. Cooler night air is first funneled through the duct, releasing several degrees of heat to the surrounding earth. As it enters the exhibit area, it is passed over an evaporative cooling fountain to increase the humidity of air, thereby cooling down the spaces. Hot air is passively exhausted at the opposite side of the building by a convection stack ventilator. With the low relative humidity range of 11-25% augmented by evaporative cooling, analysis shows that a 15°-20° temperature differential to the outside air can be expected. Cooling is further enhanced by an overhang and roll-down shades of 12% transmittance.

Electricity for lights, power and equipment (such as motor operators for night insulation and stack ventilator louvers) will be provided by an eight kilowatt wind electric generation system. Performance is estimated at 95% with a 5% projected back-up use of a propane generator.

The building is structured to display the internal workings of its natural energy systems. Items like the evaporative cooling fountain, south glass and masonry heat storage, natural ventilation systems and wind electric generator are part of the experience of visiting this park. After viewing the exhibits and the building systems, the visitor continues on the hiking trail to walk among the wildflowers, while in the distance the building becomes an inconspicuous part of the butte.

The visitor center project will be under construction in Spring 1981 with a completion date of Fall 1981.

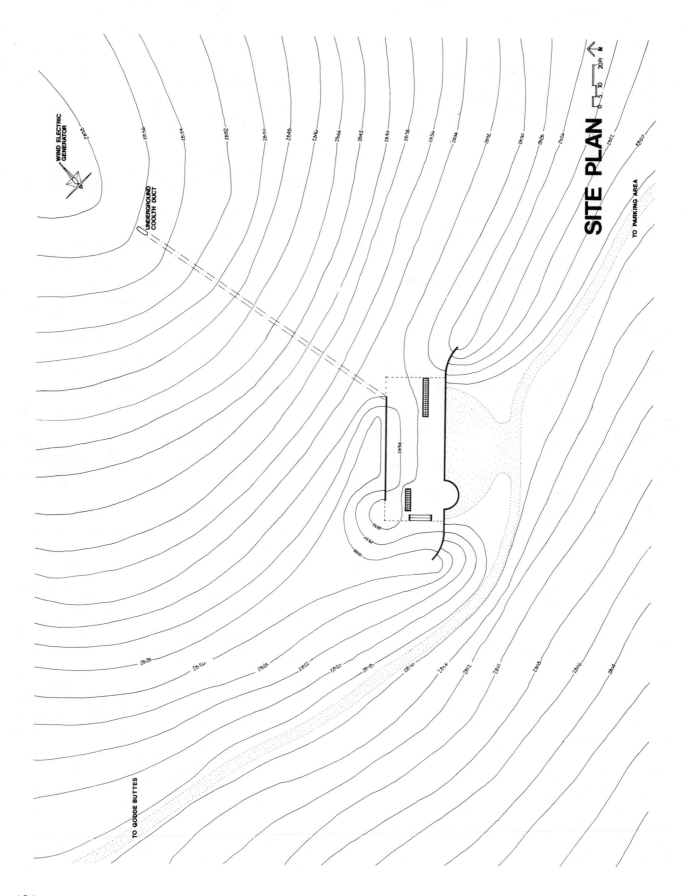

WIND ELECTRIC
GENERATOR

UNDERGROUND
COOLTH DUCT

SITE PLAN

TO PARKING AREA

TO GODDE BUTTES

0 5 10 20ft

N

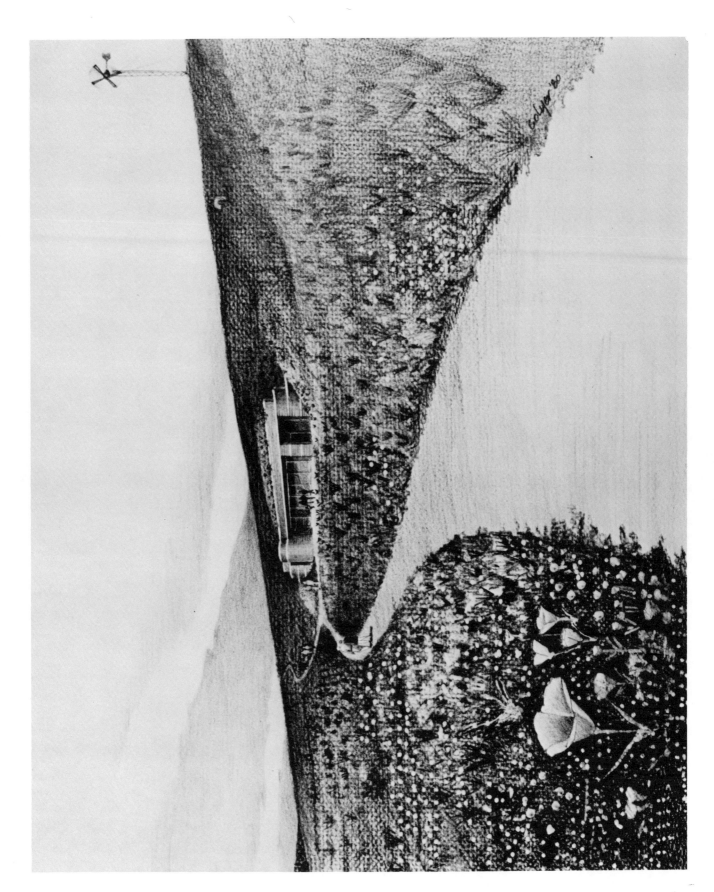

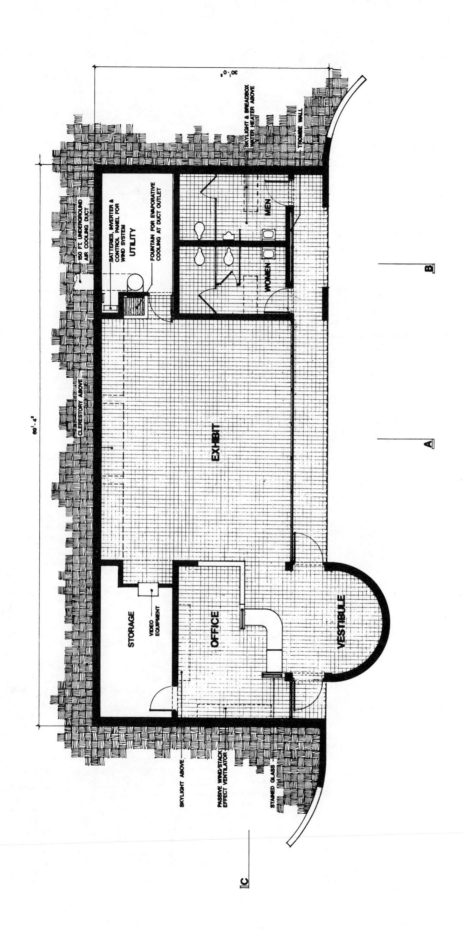

## FLOOR PLAN

0 1 2 4m

Labels within the floor plan:

30'-0"

69'-4"

SKYLIGHT & BREADBOX WATER HEATER ABOVE

TROMBE WALL

160 FT. UNDERGROUND AIR COOLING DUCT

BATTERIES, INVERTER & CONTROL PANEL FOR WIND SYSTEM

UTILITY

FOUNTAIN FOR EVAPORATIVE COOLING AT DUCT OUTLET

CLERESTORY ABOVE

MEN

WOMEN

EXHIBIT

B

A

STORAGE

VIDEO EQUIPMENT

OFFICE

VESTIBULE

SKYLIGHT ABOVE

PASSIVE WIND STACK EFFECT VENTILATOR

STAINED GLASS

C

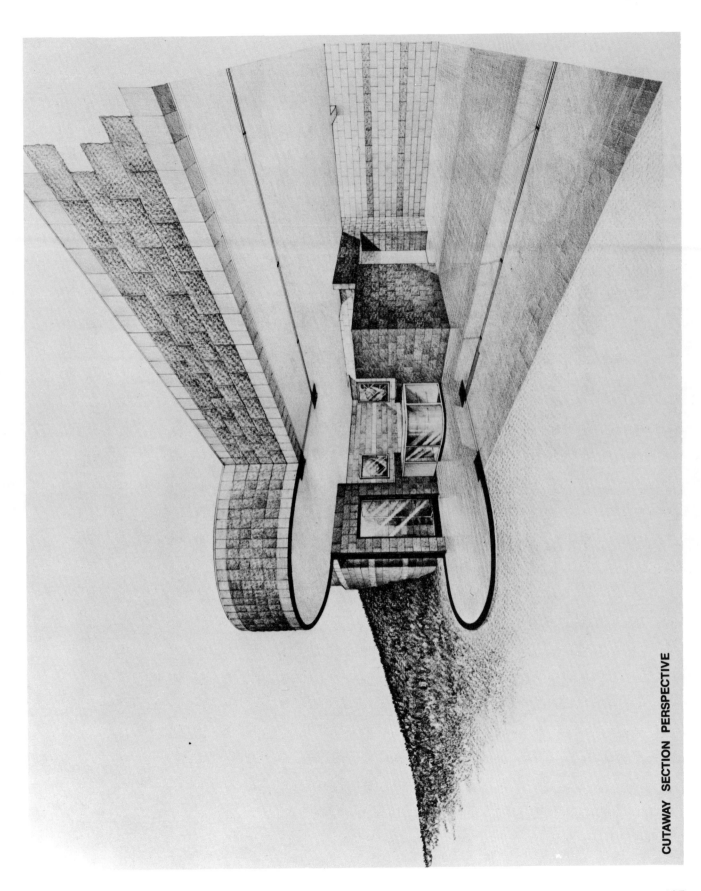

CUTAWAY SECTION PERSPECTIVE

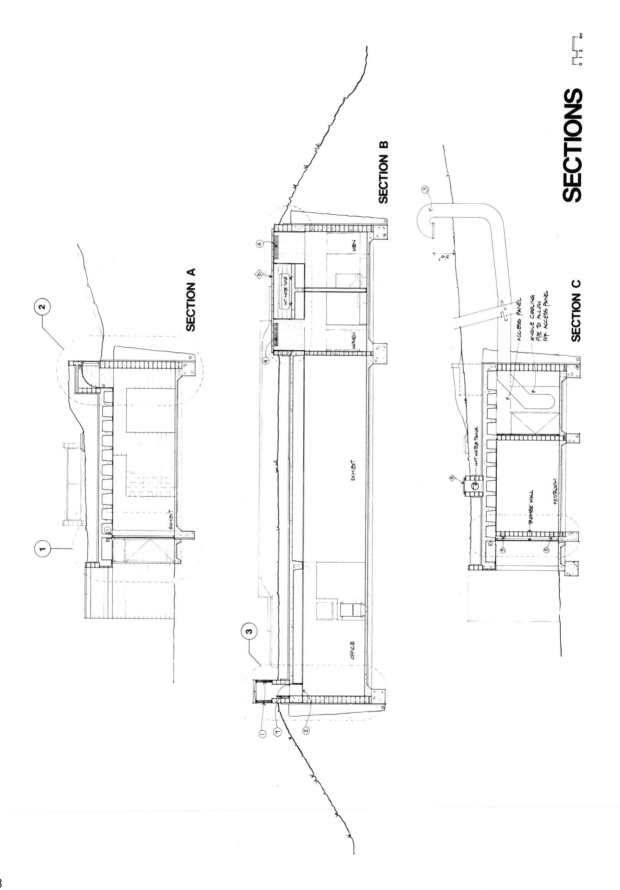

SECTION A

SECTION B

SECTION C

SECTIONS

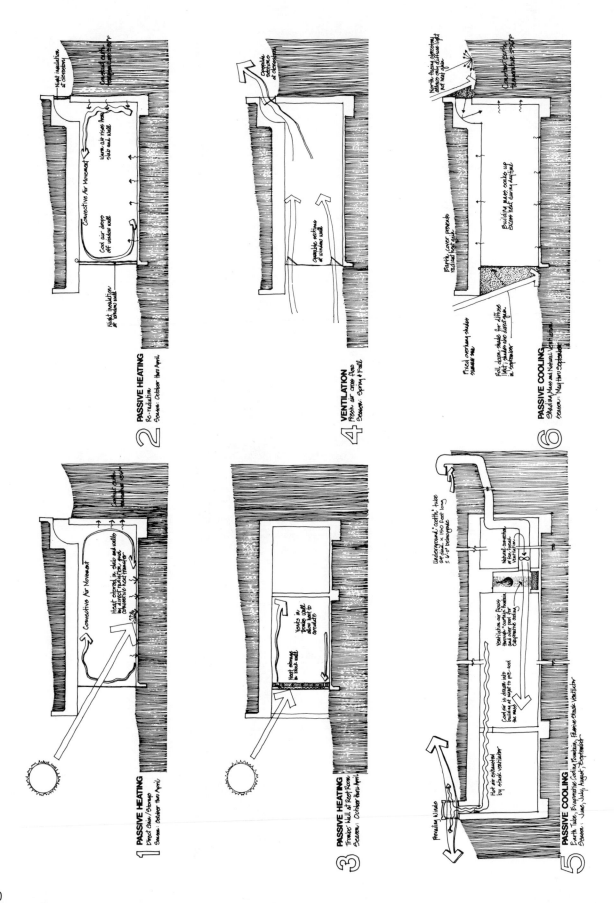

**1 PASSIVE HEATING**
Direct Gain / Storage
Season: October thru April

**2 PASSIVE HEATING**
Re-radiation
Season: October thru April

**3 PASSIVE HEATING**
Trombe Wall & Roof Pond
Season: October thru April

**4 VENTILATION**
Fresh air cross flow
Season: Spring & Fall

**5 PASSIVE COOLING**
Earth Tube, Evaporative Cooling, Fountain, Flume stack Ventilation
Season: June, July, August, September

**6 PASSIVE COOLING**
Shading, Mass and Natural Ventilation
Season: May thru September

140

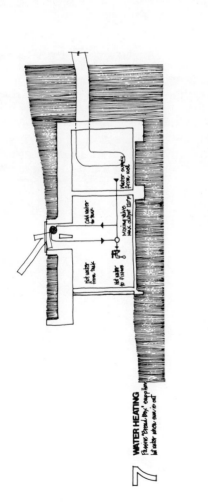

**7 WATER HEATING**

Passive "Bread Box" water heater. Let water warm in sun in day.

Hot water from tank

Hot water to fixture

Cold water to tank

Mixing valve tempers hot output temp.

Water supply from well

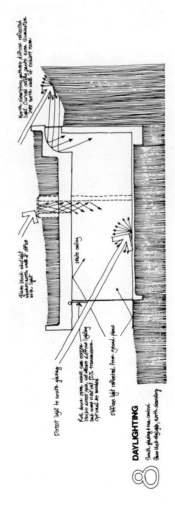

**8 DAYLIGHTING**

North clerestory gathers diffuse reflected light. Curved ceiling guides even illumination over north wall of exhibit room.

Direct light to south glazing.

Roll down open-weave sun-control blocks direct sun yet allows diffuse lighting and used with only 10% transmission. Optional to roll shade.

Diffuse light reflected from ground plane.

White ceiling

Glass block skylight. Illuminates walls & diffuse into light.

South glazing: sun control. Glass block skylight: north clerestory.

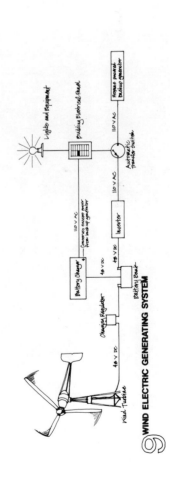

**9 WIND ELECTRIC GENERATING SYSTEM**

Lights and equipment

Building Electrical Panel

110 V AC

Automatic (AC) transfer switch

Propane powered backup generator

Battery Charger

Inverter

Converts excess power from back-up generator

110 V AC

110 V DC

48 V DC

48 V DC

Charge Regulator

Battery bank

48 V DC

Wind Turbine

# Second Award

Visitor's Center for Mesa Verde Park in
Southwestern Colorado
by Steve W. Davis, Student, Iowa State
University, Ames, Iowa

The southwest area of Colorado is a region of high deserts, out of which rise large "green tables," the Mesa Verde. Nestled in the caves and sheer canyon walls of this region are the remains of a civilization known by the Navajos of the region centuries later as the Anasazi, "the ancient ones." They left the cliff dwellings and ceremonial caves as a testament to their achievements and a source of curiosity for thousands of visitors who come to Mesa Verde Park each year.

The project was to take an area just outside of the park entrance on U.S. Highway 160 and design a visitor's information center, integrated into the landscape of Mesa Verde Park in particular, and Colorado in general. Along with providing needed information, the center should be a place of relaxation for tourists. Requirements for the center include parking for 20 cars and 20 campers, an information area, restroom and approximately ten kiosks (short term camping areas) for picnics and lunches.

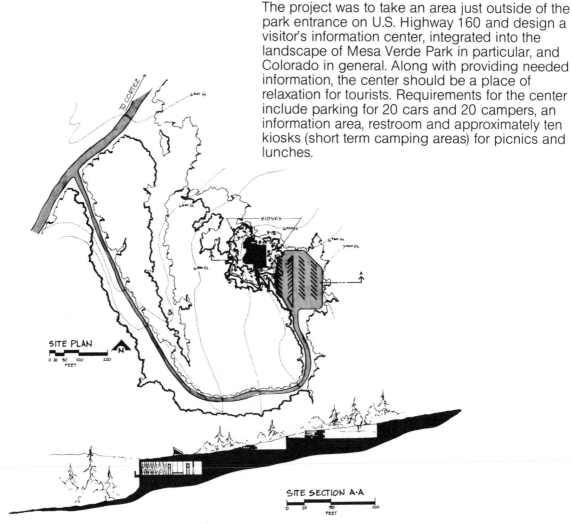

SITE PLAN

0 20 50 100 200
FEET

SITE SECTION A·A

0 20 50 100
FEET

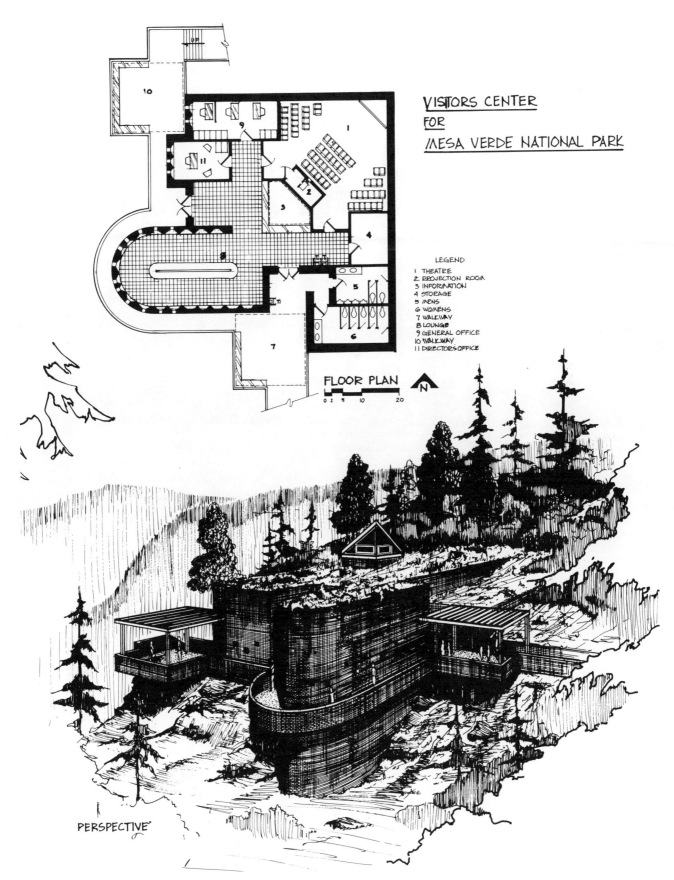

VISITORS CENTER
FOR
MESA VERDE NATIONAL PARK

LEGEND
1 THEATRE
2 PROJECTION ROOM
3 INFORMATION
4 STORAGE
5 MENS
6 WOMENS
7 WALKWAY
8 LOUNGE
9 GENERAL OFFICE
10 WALKWAY
11 DIRECTORS OFFICE

FLOOR PLAN N
0 2 5 10 20

PERSPECTIVE

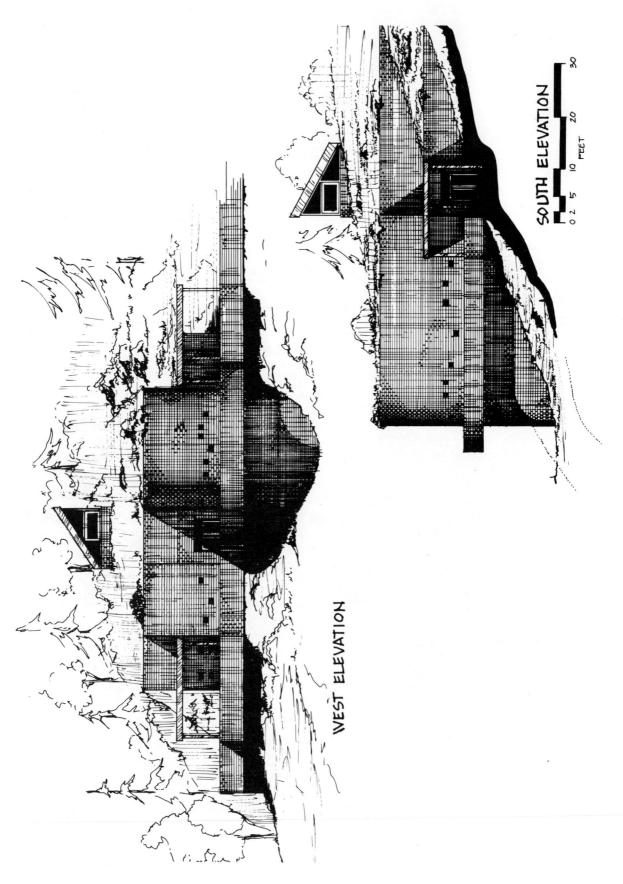

WEST ELEVATION

SOUTH ELEVATION

0 2 5    10    20    30
         FEET

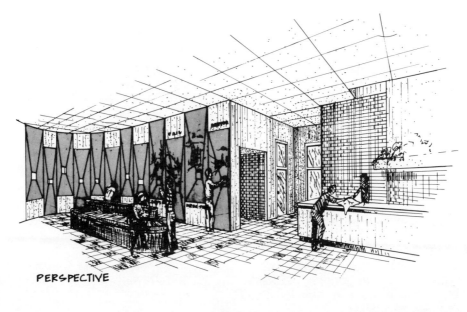

PERSPECTIVE

BUILDING SECTION

0 2 5  10        20        30
FEET

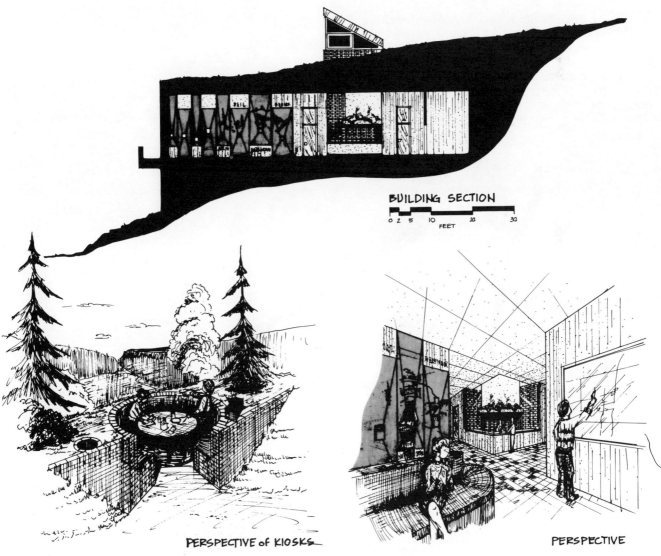

PERSPECTIVE of KIOSKS

PERSPECTIVE

## Visitor's Center for Mesa Verde Park in Southwestern Colorado
by James F. McMartin, Student, Iowa State University, Ames, Iowa

The intent of this project was to design a visitor's center for the State of Colorado along U.S. Highway 160, near the entrance to Mesa Verde National Park. The center would provide information about Colorado in general and Mesa Verde in particular, as well as restroom facilities and lounge areas for travelers.

A study of the architecture of the Mesa Verde Indians, spanning the period from 1 AD to 1300 AD, reveals that they made extensive use of many ideas that are now being rediscovered by those interested in earth-sheltered design. They built their homes in the many natural caves in the south side of the cliffs in the area and used thick adobe walls to help maintain more constant temperature in the structures.

A site, on the south side of the ridge overlooking the road and facing toward the park entrance, was selected following the examples set by the Indians. The design incorporates an entrance from behind and above. Entrance to the structure is down a circular ramp which looks out at the spectacular view. The open, central information area allows for flexibility in display and opens into a veranda which, also, has a view of the valley below.

The center is built into the hillside on three sides to take advantage of the shelter of the earth. The south elevation is an alternating series of glazing and thick pilasters, the glazing to allow the sun and the view in and the thick walls to help moderate the inside temperatures in both winter and summer.

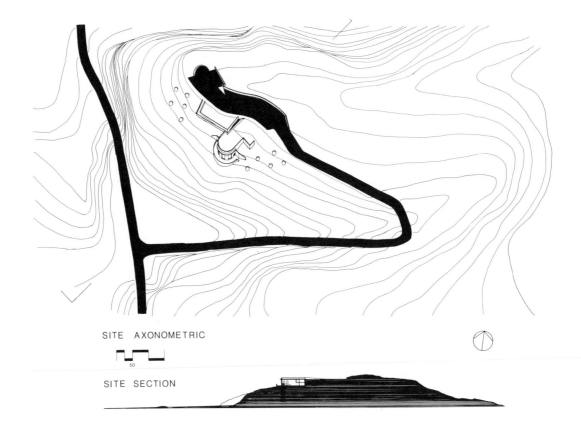

SITE AXONOMETRIC

50

SITE SECTION

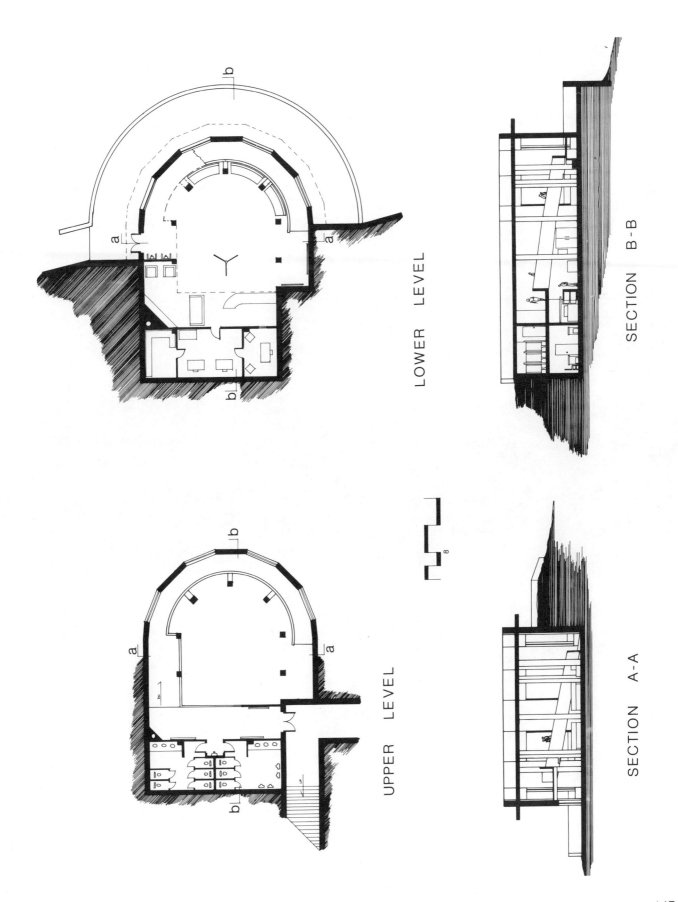

LOWER LEVEL

SECTION B-B

UPPER LEVEL

SECTION A-A

147

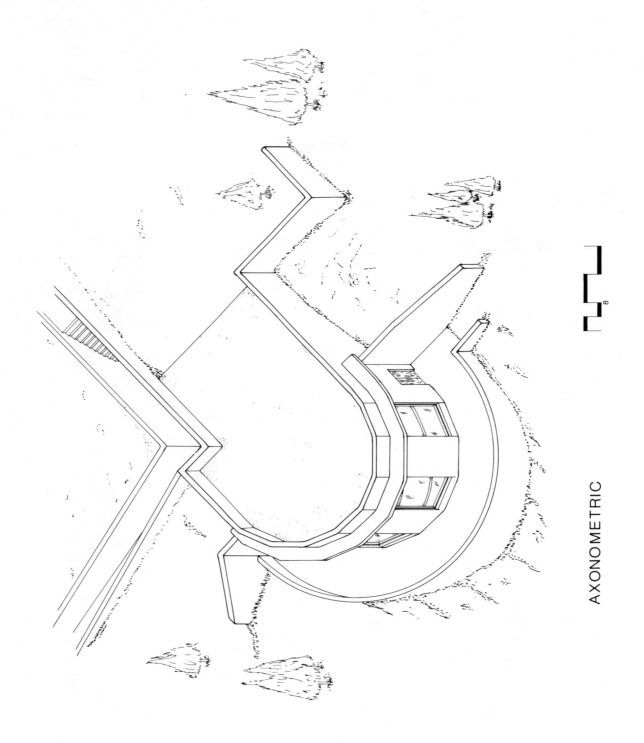

AXONOMETRIC

# Honorable Mention

## Catholic Church in a San Francisco Suburb
by Scott M. Pozzi, Student, Johnson/Anderson Architects, Palo Alto, California

Based on the actual needs and requirements of a Roman Catholic parish in the San Francisco Bay Area, this hypothetical project is a response to that parish's particular contextual characteristics as well as a reflection and embodiment of the Roman Catholic Liturgy.

The decision to effect an underground solution was predicated on three primary considerations: context, symbolism and conservation.

Situated in a virtual bowl and dominated by competitive and disruptive elements, the site was surrounded by suburban confusion. To the east, there is a major freeway. On the south is the area's largest indoor shopping mall. To the west is a mixed use, professional/light industrial/multi-family residential neighborhood. To the north is the most visible edifice in the entire basin, a ten-story hospital complex with its various support facilities atop a prominent plateau. This exterior continuum of noise and activity is the antithesis of the setting necessary for an "interior," tranquil, contemplative worship space.

The Roman Catholic faith is filled with symbolism and meaning. Its Liturgy of the Mass is a state for its enactment and actualization. Contemporary theology, unlike the middle ages, no longer intends the church to be interpreted as a monument that stands apart from the community (the most supreme example being the Gothic Cathedral). Rather, its mission is that of service to the community and, therefore, architectural harmony with it. The church building must also embody the spirits of poverty and mystery. Not to be confused with impoverishment, poverty is the loving acceptance of limitations of materials, that is, a freedom from non-essentials. Mystery in the church is the seed for creative and contemplative imagery. It is the quality which distinguishes a space as being false and pretentious from one that possesses honesty and integrity.

The spatial manifestations of these themes are more clearly illustrated by Joseph Fitzger's Four Concentric Spheres of Influence in the Liturgy, in which the two innermost spheres are the sanctuary (location of the sacraments and altar) surrounded by the congregation. These are defined by the third sphere, the interior architecture of the space and enclosed by the fourth sphere, the exterior architecture in relation to its surroundings.

In keeping with the spirit of honesty and integrity, it is important that the incorporation of these ideals into environmental system design not be overlooked. Conservation, then, becomes a critical theoretical as well as practical issue. Architectural conditions can help the individual become aware that he is part of a community with the Priest and with other members of the congregation. However, the faithful cannot participate fully in the Liturgy unless the church offers an interior environment that is comfortable.

The most effective means of achieving this consistent interior comfort level is by employing nature's most efficient heat source, the sun, and its most efficient moderator, earth. The direct heat gain derived from solar radiation captured by the south facing occulus will be stored in the massive concrete floor and walls and should provide sufficient warmth in the space.

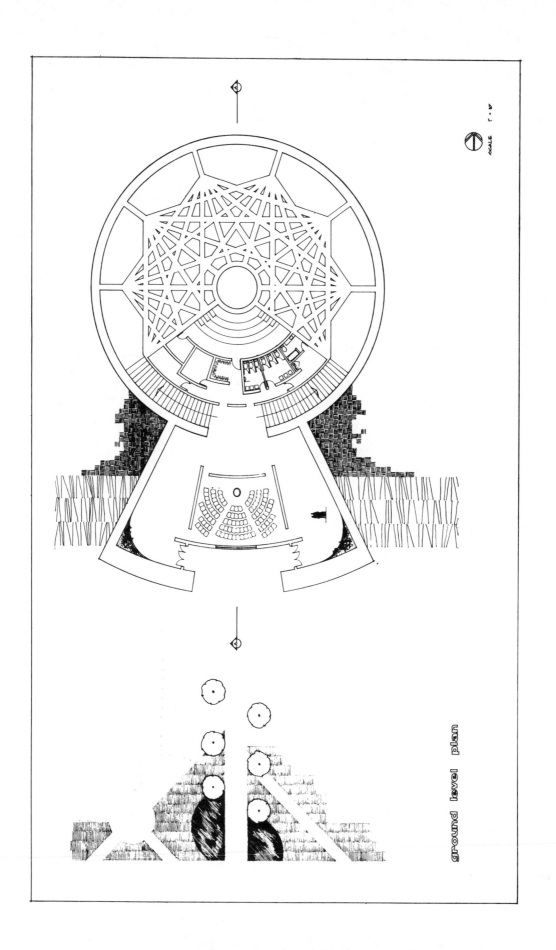

ground level plan

SCALE

150

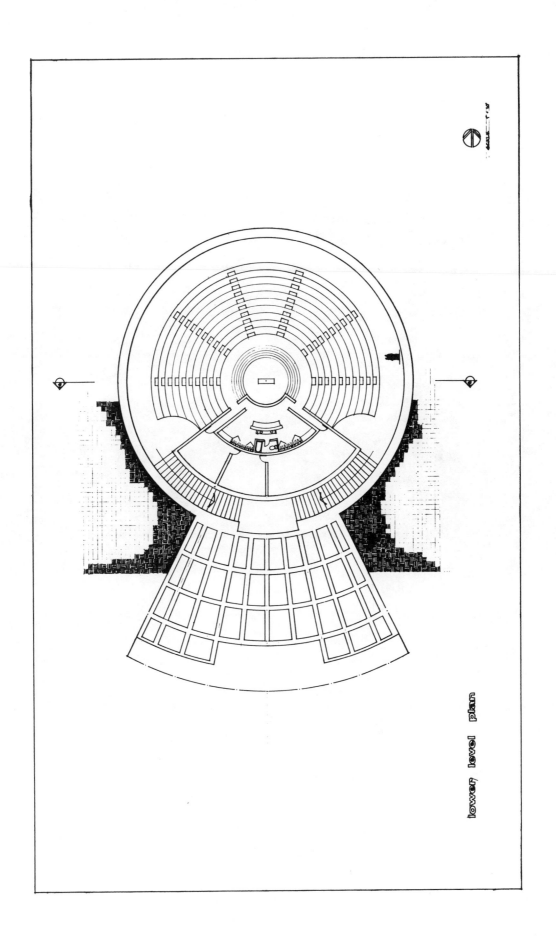

lower level plan

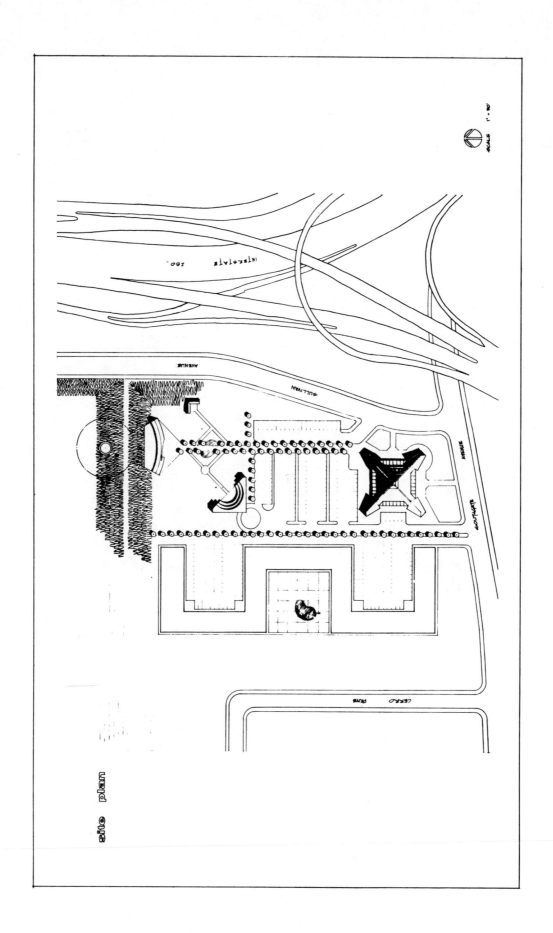

site plan

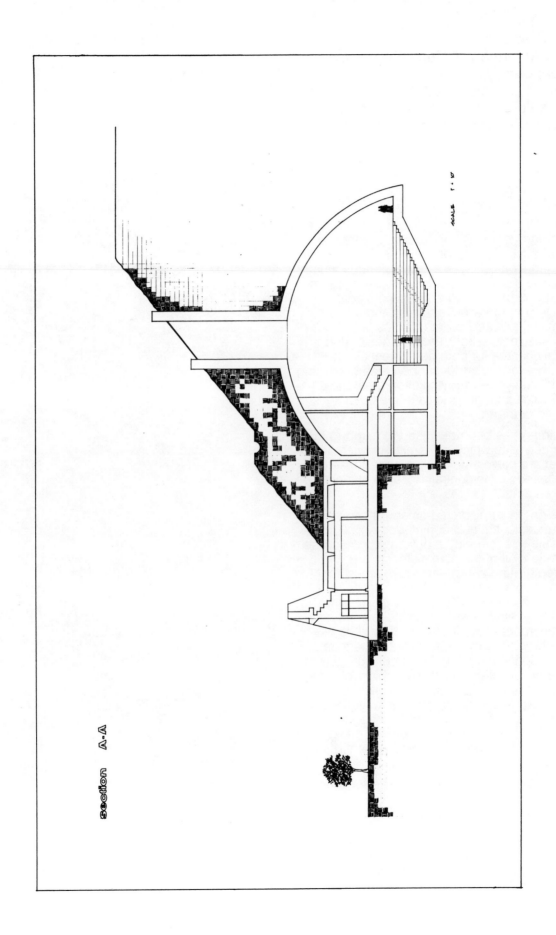

Section A-A

SCALE 1 : 10

153

## African American Cultural Center in Minneapolis, Minnesota
by Milo H. Thompson, AIA, Frederick Bentz/Milo Thompson & Associates, Inc., Minneapolis, Minnesota

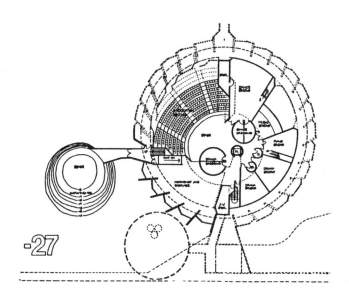

The program for this facility consists of the following basic parts: a performing arts component consisting of a 300 seat auditorium and special studio spaces for drama, dance and music; a visual arts facility including studio spaces for photography, graphics, three dimensional arts and private studios for classrooms, lecture hall, resource center and a large gallery area for both permanent and temporary exhibits; and a support area consisting of administrative offices, workshop area and a cafeteria. The total gross area of the building is 45,000 square feet.

The building has been designed to respond to any of three sites which are similar in orientation and offer the basic amenity of a park-like setting. The preferred site is centrally located in the downtown area.

The design concept, based on an African-American historical and contemporary cultural point of view, is "centric" in building form and plan organization. All of the plan elements are organized around a large atrium space which functions as a meeting place and a focus for all of the activity contained in the building. A large 100 foot diameter fiberglass skylight covers the main exhibition spaces and the atrium. It and other interior skylights allow natural light to penetrate the entire space including the circulation reas at the lower levels of the building. A solar collector, designed to feature its technical, aesthetic and symbolic properties, is arranged as a large scale venetian blind over the skylight to shade it from direct sunlight. The collector system is designed to provide 60% of the energy requirement of the building.

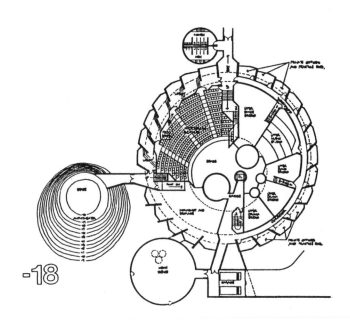

The project is designed as a long-life, low-energy-consuming, low-maintenance building. Exploiting the geometric properties of a conical form with the base set into the earth, approximately half of the building area and volume is beneath grade, utilizing the earth's natural moderating qualities. The building will be set into a flat site at the mid-level of its section, and will use the excavated

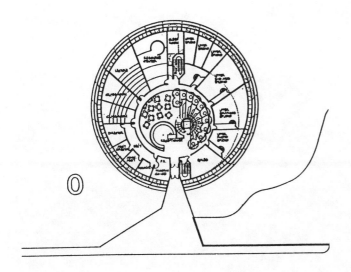

material for fill around the periphery, raising the grade elevation above the street and creating irregularly contoured landscape features at the edges of the site. The forms anticipate the location of expansion to the building, using circular courtyards depressed below grade to light the underground encompassing space. The courtyards and their related spaces would be linked together and to the main building by underground corridors.

The structure of the building consists of concrete piers and foundation walls, wide-flange steel columns for the conical form of the out-of-ground structure, and a two-way "bow string" steel truss arrangement supporting the skylighted roof. The inclining walls, finished on the exterior with metal, accommodate vertical ducts fed from a main horizontal distribution system incorporated into the concrete pier design at the grade floor elevation of the building.

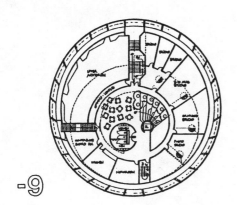

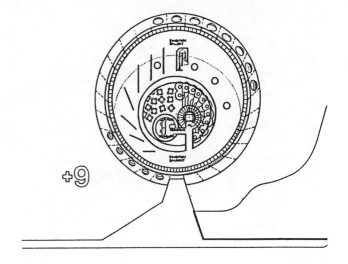

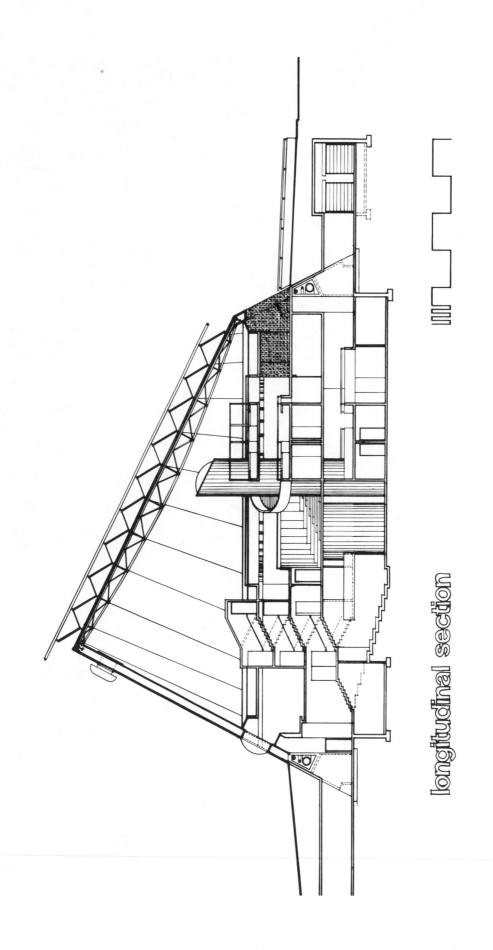

longitudinal section

# Corporate Headquarters for Solar Ray, Inc. in Peoria, Illinois
## by Ralph Johnson, Architect, Perkins & Will, Chicago, Illinois

The program is to provide a facility of approximately 30,000 square feet to house the corporate headquarters of a company which distributes energy-related products along with rentable office and warehouse space. Three types of space are to be provided: office, warehouse and a showroom for product display. The client requested that the building solution incorporate ideas using passive as well as active energy concepts.

A two-story office area is placed to the north along with an axial circulation spine. The southern portion contains a one-story warehouse and showroom area. The overall form of the building is stepped in plan to produce a maximum wall exposure to the south. This south wall incorporates a trombe wall adjacent to warehouse spaces and a direct gain glass wall adjacent to the showroom. To the south of this wall is a solar pond and a solar field and windmill.

One of the primary goals for the Solar-Ray project was to achieve a 50% lower annual energy usage than an equivalent building incorporating standard architectural and environmental control systems. The results of this combined architectural and engineering effort is a building which has a 19% reduction in maximum heating load and a 33% reduction in maximum cooling load when compared to an equivalent building meeting the provisions of energy performance standards proposed by the United States Congress.

Solar-Ray achieves its low annual energy usage by taking maximum advantage of passive solar energy to reduce building illumination and heating requirements. The storage and insulating benefits of earth-berming are used extensively in the warehouse areas to further reduce heating and cooling loads. In addition, earth berms are used on the north side of the building to deflect winter winds. The south side of the berm is a battered wall with a reflective surface to provide additional heating and light on the north side of the building.

The heating, ventilating and air conditioning system is designed to make maximum use of the solar energy collected by the trombe wall and the direct gain wall. During mild weather, the ventilation system distributes the air warmed by the sun to those areas requiring heating. During cold weather, the excess solar energy received is stored for use at night through the use of a heat pump.

Artificial illumination requirements are held to a minimum by the use of large, well-shaded glass areas. Insulated louvers incorporated into the air space between the glass areas are closed at night providing a heat loss equivalent to that for an insulated wall surface. In addition, masonry and concrete bearing walls with steel structural frame are used.

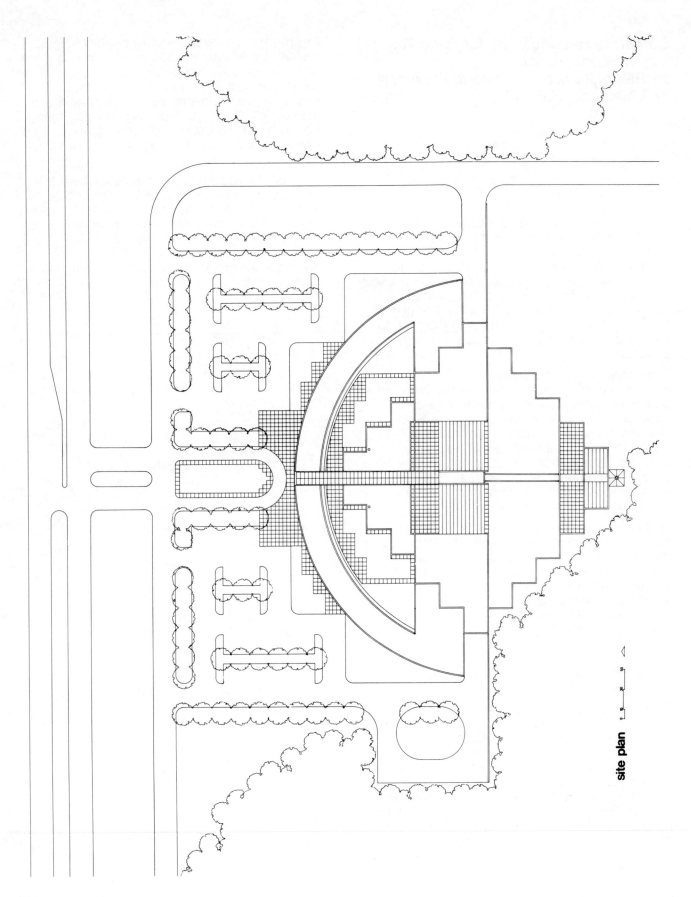

**site plan**

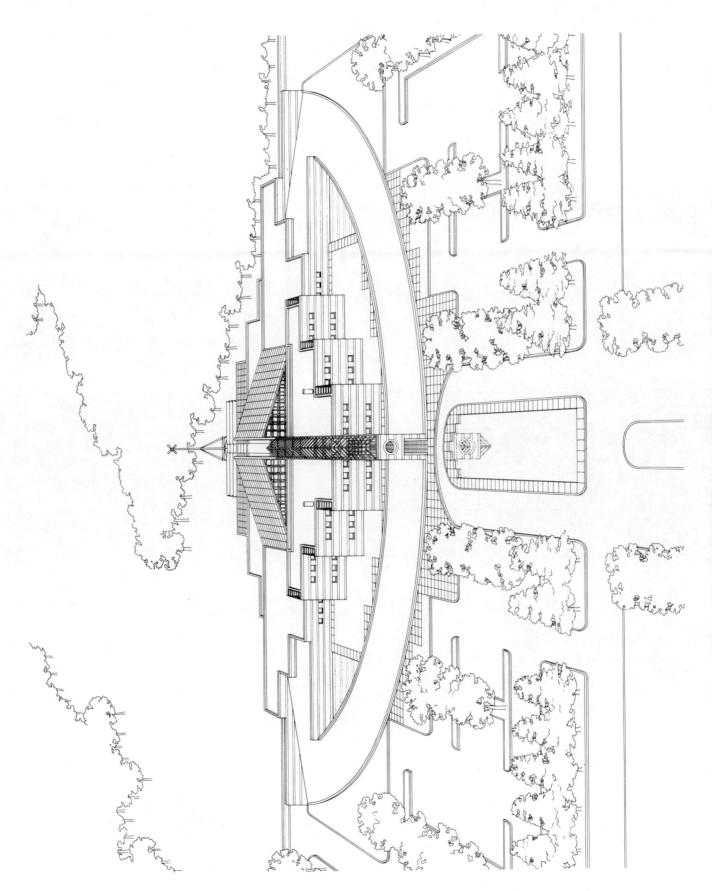

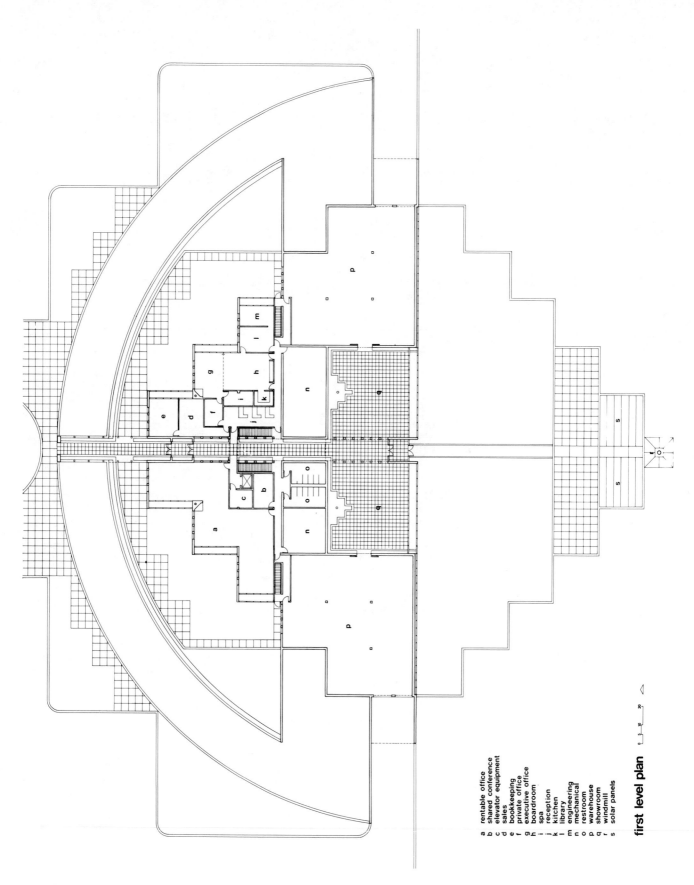

a rentable office
b shared conference
c elevator equipment
d sales
e bookkeeping
f private office
g executive office
h boardroom
i spa
j reception
k kitchen
l library
m engineering
o restroom
p warehouse
q showroom
r windmill
s solar panels

**first level plan**

0  5  10  20

162

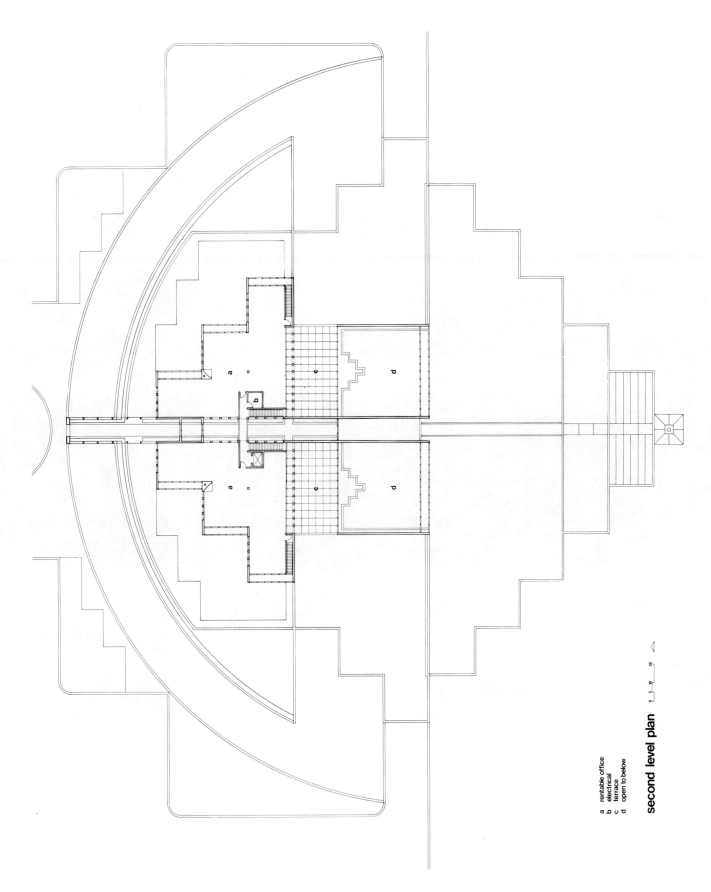

**second level plan**

a   rentable office
b   electrical
c   terrace
d   open to below

0  5  10    20

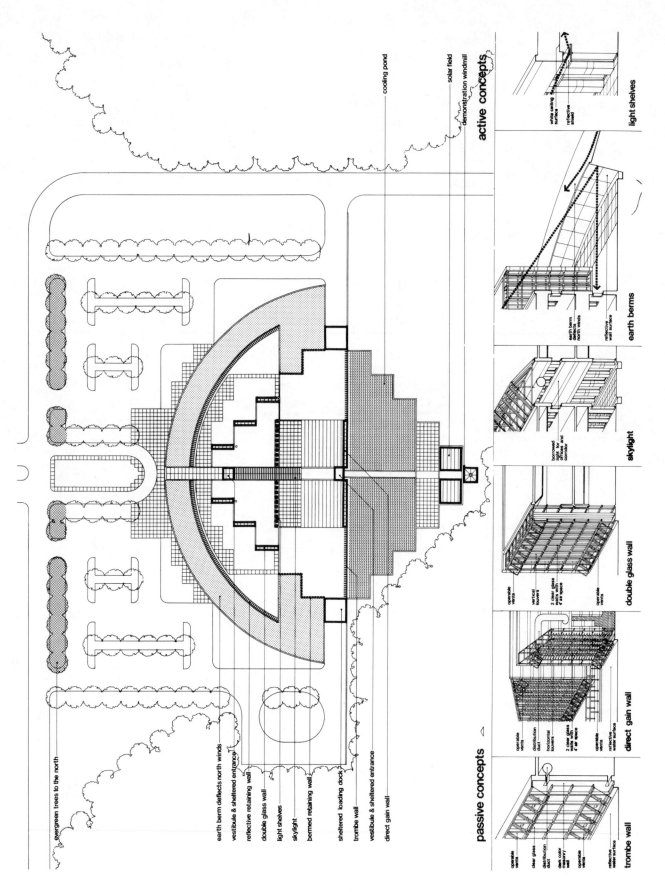

evergreen trees to the north

earth berm deflects north winds
vestibule & sheltered entrance
reflective retaining wall
double glass wall
light shelves
skylight
bermed retaining wall
sheltered loading dock
trombe wall
vestibule & sheltered entrance
direct gain wall

cooling pond

solar field
demonstration windmill

**active concepts**

**light shelves**

white ceiling surface
reflective shield

**earth berms**

earth berm deflects north winds
reflective wall surface

**skylight**

borrowed light for offices and corridor

**double glass wall**

operable vents
vertical louvers
2 clear glass walls with 4' air space
operable vents

**direct gain wall**

operable vents
distribution duct
horizontal louvers
2 clear glass walls with 4' air space
operable vents
reflective water surface

**trombe wall**

operable vents
clear glass
distribution duct
dark color memory wall
operable vents
reflective water surface

**passive concepts**

164

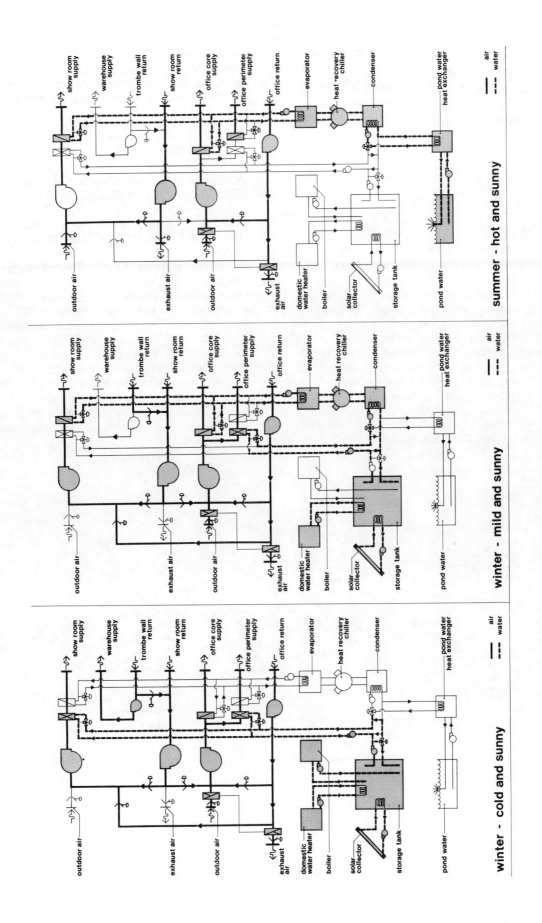

summer - hot and sunny

winter - mild and sunny

winter - cold and sunny

## Office Industrial Complex for Westville, New Jersey
by Ronald Kardon and Michael Kleintop, Architects, Francis G. Vitetta Associates, Westville, New Jersey

The complex consists of a 68,000 square foot, single-level building containing offices, shops and warehouse space and a 40,000 square foot service area also containing shops and interior and exterior storage facilities. Because of the surrounding existing structures, vegetation, lines of sight from existing structures and site grading conditions, the location of the underground building was a program requirement. In addition, the structural concrete warehouse and shops will have activities requiring a small temperature change and little heat loss or gain.

The main structure is situated at the service road level, excavated into the retained earth which has an average height of 18 feet above the service road. Shops and storage facilities are earth-covered, while the office area is roofed with a system of skylights and solar collectors.

The offices acquire heat gains during the work day and these necessitate greater temperature control. Ducts running six feet below grade northwest of the warehouse will channel fresh air into the building, augmenting space-heating and cooling requirements throughout. The skylighted office core is steel frame construction, allowing for proper heat loss through the graveled roof. Well water on site will be utilized to chill the ducted air into these spaces during the cooling season. Space heating is accomplished with the direct-gain passive solar southwest wall, with glazing recessed from the building line. Reflective exterior louvers modulate solar radiation absorption through these windows and the southwest-facing skylights over the service corridor. Flat plate solar collectors located at the north perimeter of the office core and above the storage sheds at the service area compound satisfy domestic hot water and additional heating requirements. Existing and new vegetation provide additional screening and seasonal shading.

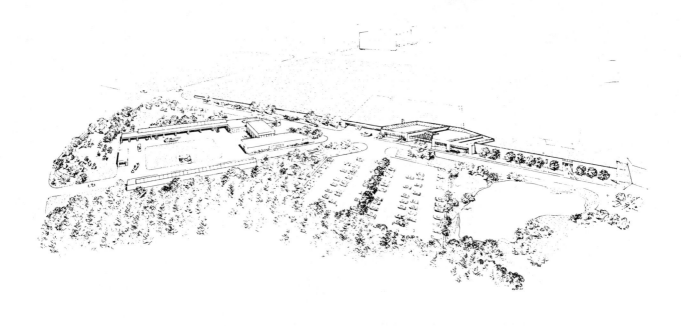

VIEW FROM SOUTH

2-15-81

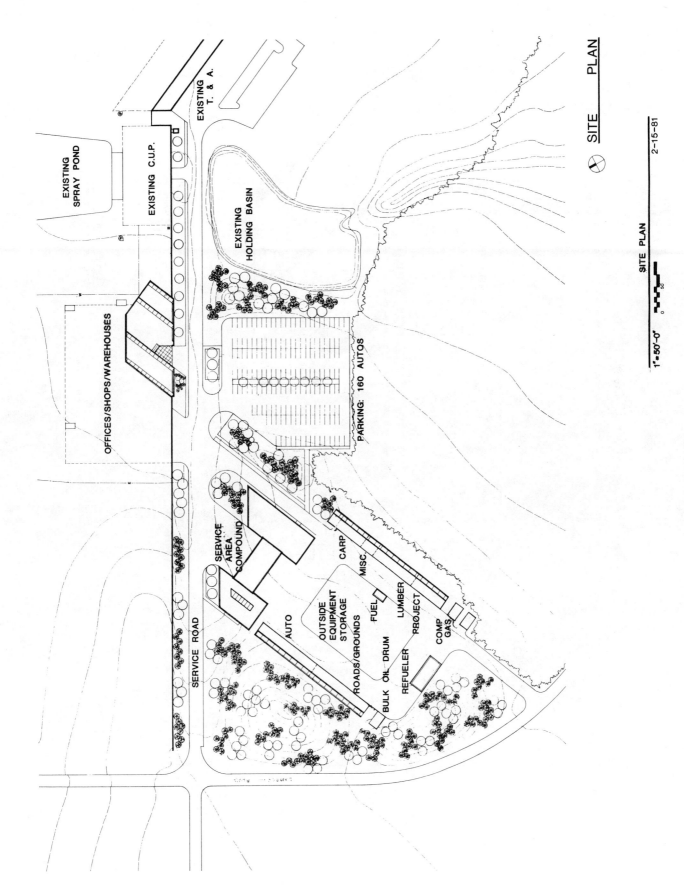

EXISTING SPRAY POND

EXISTING C.U.P.

EXISTING T. & A.

EXISTING HOLDING BASIN

OFFICES/SHOPS/WAREHOUSES

PARKING: 160 AUTOS

SERVICE ROAD

SERVICE AREA COMPOUND

AUTO

OUTSIDE EQUIPMENT STORAGE

ROADS/GROUNDS

BULK OIL DRUM

REFUELER

FUEL

LUMBER

PROJECT

COMP GAS

CARP

MISC

SITE PLAN

SITE PLAN

1"=50'-0'

0    50

2-15-81

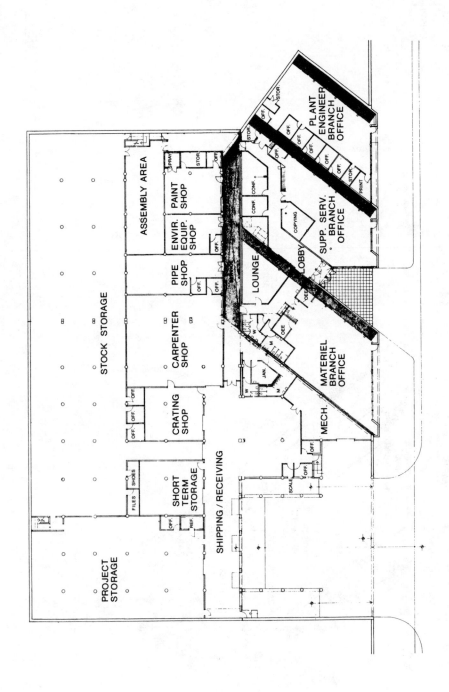

OFFICES/SHOPS/WAREHOUSES: **FLOOR PLANS**

1/16" 1'-0"    2-15-81

PROJECT STORAGE

STOCK STORAGE

ASSEMBLY AREA

SHORT TERM STORAGE

CRATING SHOP

CARPENTER SHOP

PIPE SHOP

ENVIR. EQUIP. SHOP

PAINT SHOP

SHIPPING / RECEIVING

FILES

SHOES

OFF.

REF.

OFF.

OFF.

OFF.

OFF.

SPRAY

STOR.

STOR.

OFF.

OFF.

OFF.

LOUNGE

CONF.

CONF.

STOR.

STOR.

OFF.

OFF.

OFF.

OFF.

OFF.

PRINT

STOR.

STOR.

PLANT ENGINEER BRANCH OFFICE

SUPP. SERV. BRANCH OFFICE

COPYING

LOBBY

OFF.

OFF.

MATERIEL BRANCH OFFICE

W

M

W

M

JAN.

OFF.

MECH.

SCALE

OFF.

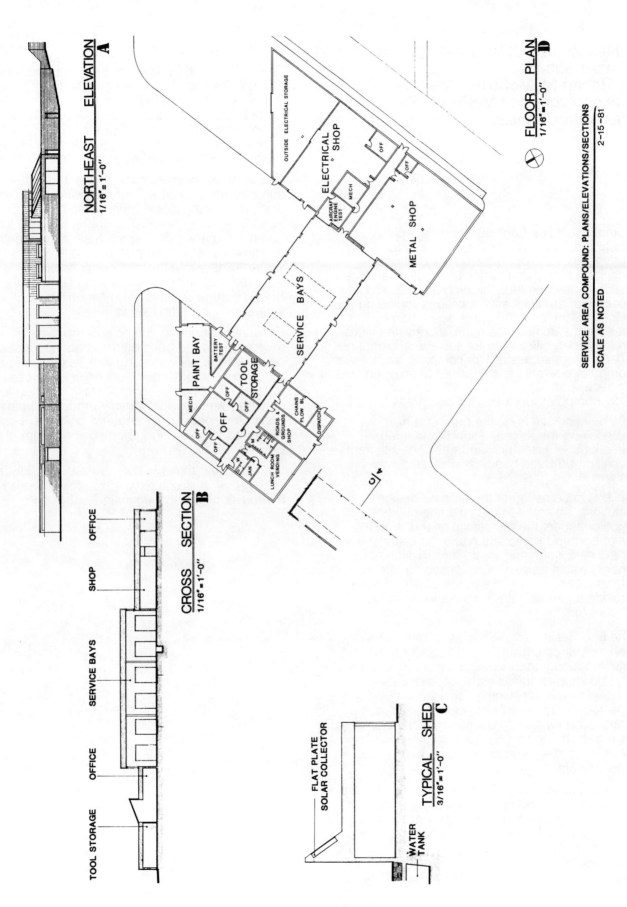

NORTHEAST ELEVATION **A**
1/16"=1'-0"

CROSS SECTION **B**
1/16"=1'-0"

TOOL STORAGE OFFICE SERVICE BAYS OFFICE SHOP OFFICE

FLAT PLATE SOLAR COLLECTOR

WATER TANK

TYPICAL SHED **C**
3/16"=1'-0"

OUTSIDE ELECTRICAL STORAGE

ELECTRICAL SHOP

OFF

MECH

AIRCRAFT ENGINE TEST

OFF

METAL SHOP

SERVICE BAYS

PAINT BAY

BATTERY TEST

MECH

OFF

OFF

OFF

TOOL STORAGE

OFF

ROADS & GROUNDS SHOP

CHAINS PLOW BL

DISPATCH

LUNCH ROOM VENDING

JAN

FLOOR PLAN **D**
1/16"=1'-0"

SERVICE AREA COMPOUND: PLANS/ELEVATIONS/SECTIONS

SCALE AS NOTED

2-15-81

169

# Administrative Office Building for a Kansas Suburb
by Robert M. McCulley, AIA,
Wolfenbarger and McCulley, PA,
Manhattan, Kansas

Communications Services, Inc. is a rapidly growing operator of cable television and other communications outlets. Due to expanding business, a new and larger administrative office building was needed. The program required spaces for administrative functions, including offices for the company president and vice president, a conference room, six private offices and an open office area for secretarial personnel. Space was also desired for company finance and computer operations, comprised of an open area for secretaries, three private offices, a keypunch area, computer room and storage areas.

The building site is on the brow of a hill overlooking the Kansas Flint Hills in three directions — east, north and west. The site forms a transition between business and residential zoning.

Several cncerns led to the ultimate design solution. The architects felt that the building should respect the residential character of the neighborhood to the south and the distant views to the east, north and west. Conservation of energy was a primary consideration in the design of this facility, as was the desire of the owner to provide a pleasant environment for employees and visitors.

The final design solution is an earth-sheltered, concrete structure enclosing 11,300 square feet, situated along and following the shape of the hill. The building is buried eight feet along its southern exposure, which consists of a vine covered, wooden trellis structure that shades a continuous skylight, presenting a non-commercial face to its neighbors. Stone, concrete, wood and plant materials blend the building into its hilltop environment.

Energy conservation has been achieved through a combination of passive solar features, building mass, heavy insulation created by earth-sheltering and an innovative heating and cooling system. The continuous south-facing skylight allows winter sun into the building during all daylight hours. The heat from this sunlight is collected and stored in the building's concrete mass.

The mechanical system consists of a series of internal water-to-air heat pumps distributed throughout the building. This allows for transfer of excess heat from the skylight spaces to areas where heat is needed. In times when excess heat is produced, it is stored in a 5,000 gallon water storage tank buried to the south of the building.

The wood trellis shades the skylight from the summer sun and allows running vines, planted on the roof, to hang down, thereby increasing shade, as well as visually breaking up the building's form. The mass of the building will have a dampening effect on extreme temperatures, both summer and winter. To facilitate this, the concrete ceiling is left exposed and finished with a textured coating. Sound control is achieved by acoustical baffles suspended from the ceiling.

For summer cooling, the heat pump cycle is reversed, and the building mass and storage tank are utilized to allow cooling for peak loads to be produced at night when better efficiency can be obtained. Lighting is generally task oriented to conserve energy. All internally produced heat from computers, people, lighting and equipment can be stored for use during non-solar heating periods.

It is estimated that this building will use 25% less energy than a structure which is conventionally heated, cooled and illuminated.

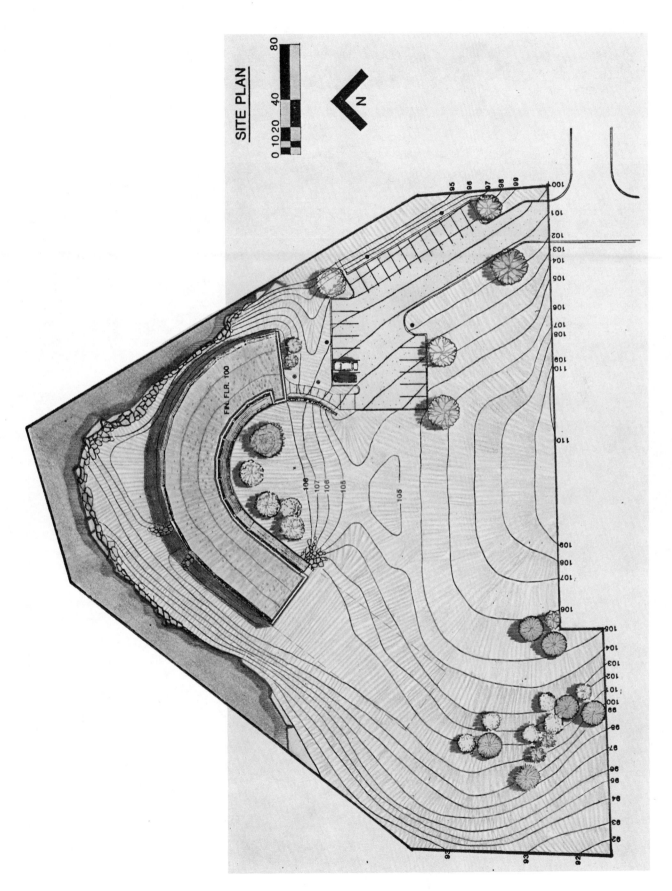

SITE PLAN

0 10 20 40 80

N

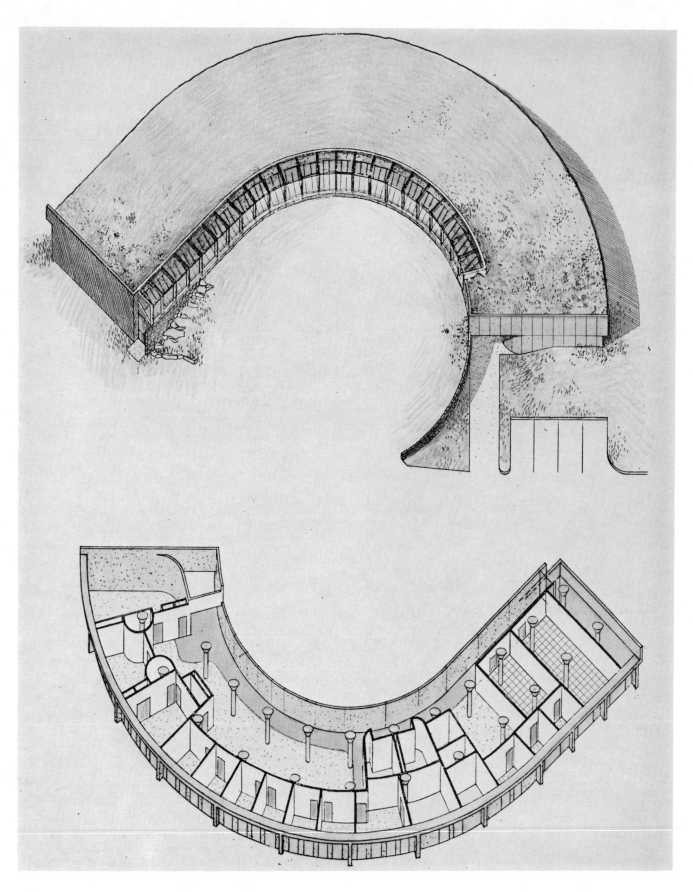

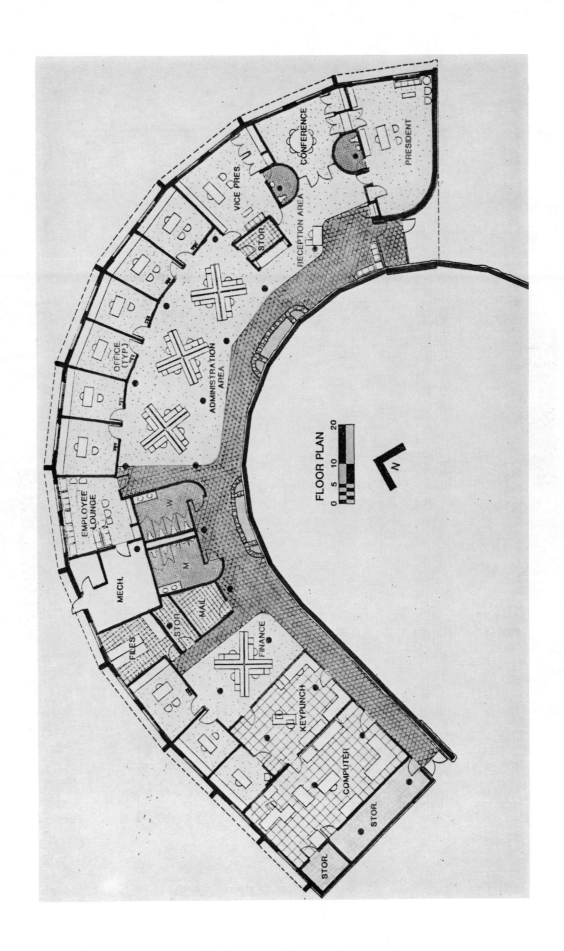

FLOOR PLAN

0  5  10  20

N

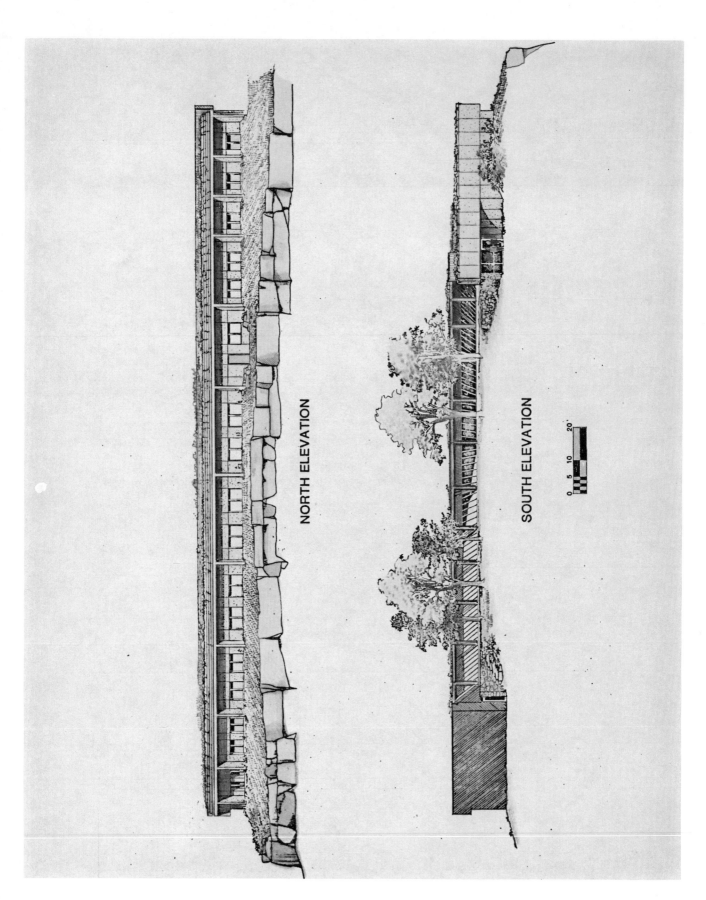

NORTH ELEVATION

SOUTH ELEVATION

0 5 10 20'

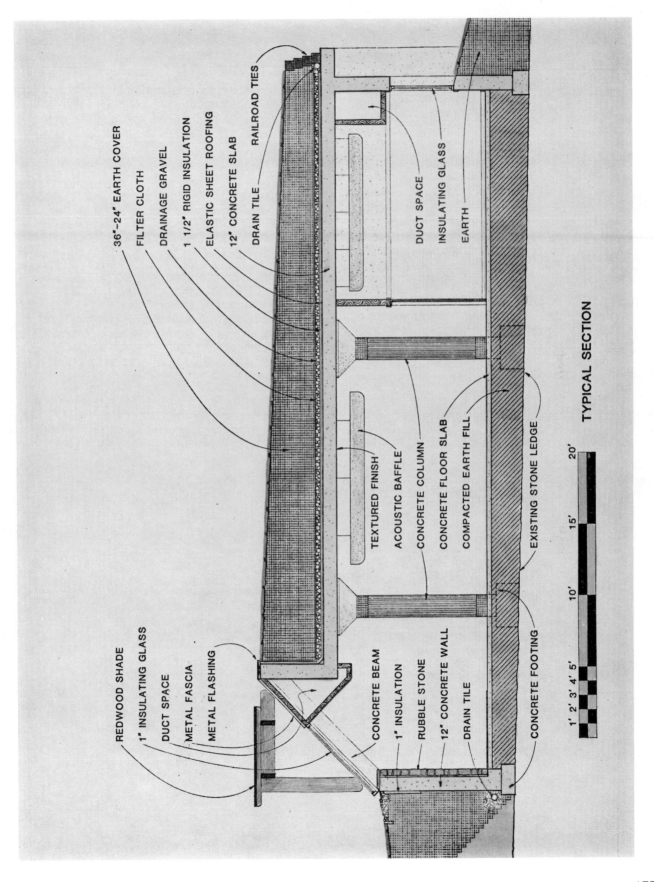

REDWOOD SHADE

1" INSULATING GLASS

DUCT SPACE

METAL FASCIA

METAL FLASHING

CONCRETE BEAM

1" INSULATION

RUBBLE STONE

12" CONCRETE WALL

DRAIN TILE

CONCRETE FOOTING

36"–24" EARTH COVER

FILTER CLOTH

DRAINAGE GRAVEL

1 1/2" RIGID INSULATION

ELASTIC SHEET ROOFING

12" CONCRETE SLAB

DRAIN TILE

RAILROAD TIES

DUCT SPACE

INSULATING GLASS

EARTH

TEXTURED FINISH

ACOUSTIC BAFFLE

CONCRETE COLUMN

CONCRETE FLOOR SLAB

COMPACTED EARTH FILL

EXISTING STONE LEDGE

TYPICAL SECTION

1' 2' 3' 4' 5'        10'        15'        20'

## Child Care Center for an Air Force Base in San Antonio, Texas
by Larry O'Neill, Architect, O'Neill, Perez and Associates, San Antonio, Texas

The architect was charged with the responsibility of providing a child care center of approximately 7,140 square feet for Brooks Air Force Base in San Antonio, Texas. The support facilities included parking, playground equipment, sidewalks and a covered loading and unloading area.

The primary function consists of administration, kitchen and multi-purpose areas and rooms for four groups of children: pre-school age three to six years, toddlers age 18 months to three years, infants age six to 18 months and nursery age six weeks to six months.

Basic design criteria consisted of providing an energy efficient building which related

architecturally to its principal users, the children. Also important was the concern that this facility not have the scale and proportion of an institutional facilitiy.

The site which slopes significantly to the south is a perfect candidate for an earth-sheltered structure. The energy savings and minimum exterior maintenance associated with earth-sheltered buildings is additional encouragement for its use.

In an effort to leave as much of the site as open as possible for play areas, the building is located where the ground is steeper, which better suits earth-sheltered structures. The higher elevation affords maximum benefit from the prevailing southeast breeze.

Interior circulation is combined to a single element made purposely wider than a normal hall. It can then act as a gallery for display as well as a play area during inclement weather. This circulation spine also serves as a gathering area for parents during peak drop-off and pick-up times and as added space for the multi-purpose dining area.

The required three-car, covered, child delivery structure is placed parallel to the contours to

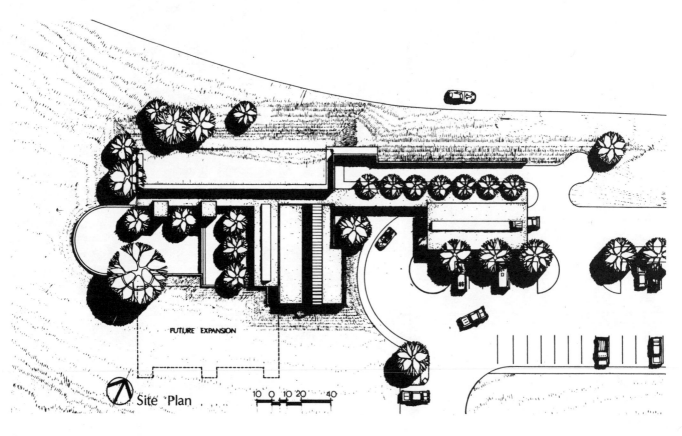

FUTURE EXPANSION

Site Plan          10 0 10 20    40

avoid the possibility of unattended cars rolling. This element is used to further reinforce the strong circulation concept.

The exterior surface material is eight square inches of brown clay tile. The use of this tile assures minimum maintenance and visual compatibility with existing structures. The brown clay tile is also a warm inviting tone appealing to the children.

The south exposure excludes the hot summer sun with a two foot overhang but allows the lower winter sun to enter. The earth cover on the roof, although not possessing a high U value, is a significant moderator of changing temperatures. At a depth of six feet the earth's temperature is relatively constant at about 70°F. Thus the earth below and around the structure acts as a thermal mass constantly offsetting the seasonal temperature extremes of the rooms above.

San Antonio has many days in the spring and fall when air-conditioning is not necessary. This structure is designed and oriented to take full advantage of cross ventilation of natural breezes by using maximum opening operable windows and exhaust monitors. In addition, future installation of solar power systems for domestic hot water and air-conditioning needs will be provided for.

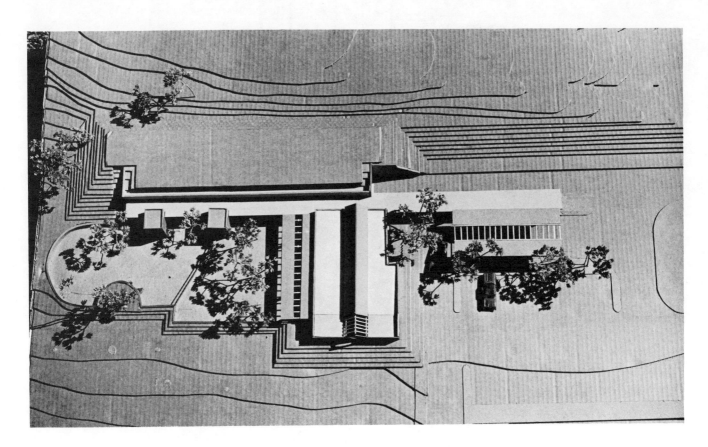

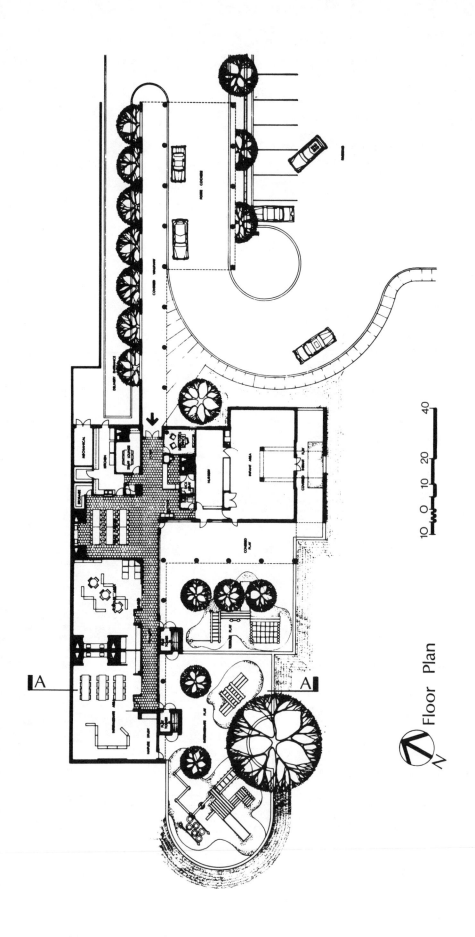

Floor Plan

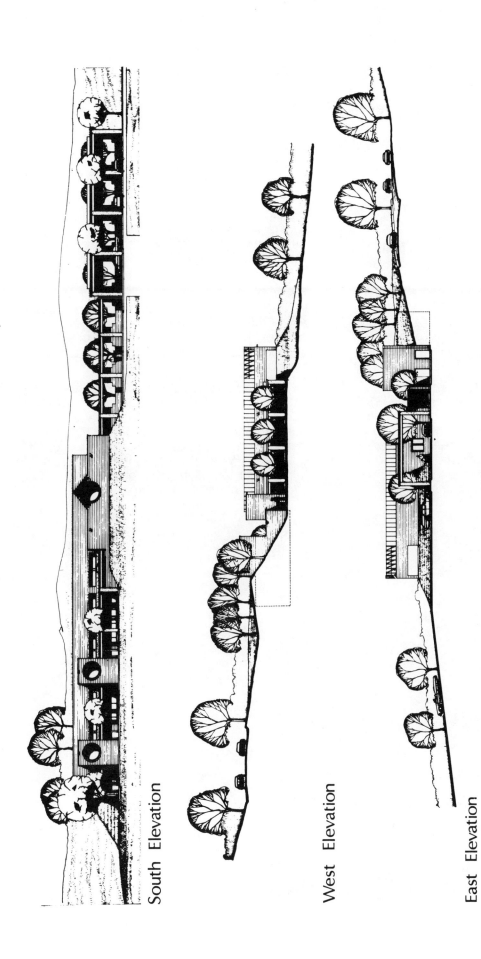

South Elevation

West Elevation

East Elevation

Section A·A

# Merit Award

## Montessori School and Administrative Building in St. Paul, Minnesota
### by Ralph Rapson, President, Ralph Rapson & Associates, Inc., Minneapolis, Minnesota

This is a design for an energy conscious and economic preschool and elementary school with additional facilities to house teacher training program and administrative needs of the Montessori Foundation of St. Paul, Minnesota. Basic to the charge was the design of a structure embodying the educational and philosophical concepts of the Montessori Foundation and a design closely related to the environment.

The Montessori approach aims to insure the normal development of the whole personality of the child — the physical and emotional facilities as well as the intellectual powers. This learning process is attuned to the child's inner drives and promotes the beneficial interaction of the child and his surroundings leading to the child's mastery of himself and his environment.

The plan uses energy conscious design techniques to create a comfortable, safe, inspiring and serene child-oriented facility embodying the Montessori philosophy. It also integrates internal with external use and is nature oriented in an urban setting.

The urban site, in a moderate income level residential area, is bounded on the north by a major freeway frontage road and on the south by a residential street. A small park with a Community Center, a Head Start School and residences are immediate neighbors. The site falls 25 feet from the south to the north.

The design solution places internal functions below grade with the classrooms generally to the south and the offices, teacher-training facilities, etc. at the north. Pedestrian entrance at upper ground level is from the south with bus and vehicle entrance from the lower frontage road at the north.

All classrooms have sunken, south-oriented, exterior activity courts and immediate access to central, common space and other active and passive functions. A system of roof monitors provide flat plate solar collectors and passive solar glass monitors. A greenhouse and an open internal garden court provide functional integration of nature and structure.

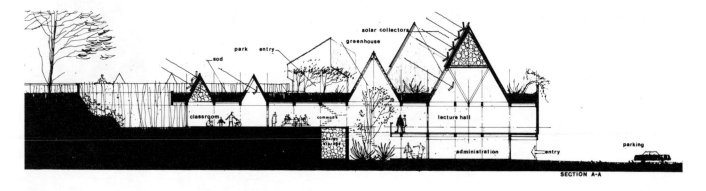

SECTION A-A

MONTESSORI SCHOOL & ADMINISTRATIVE BUILDING, MONTESSORI FOUNDATION OF MINNESOTA INC, ST PAUL MINNESOTA

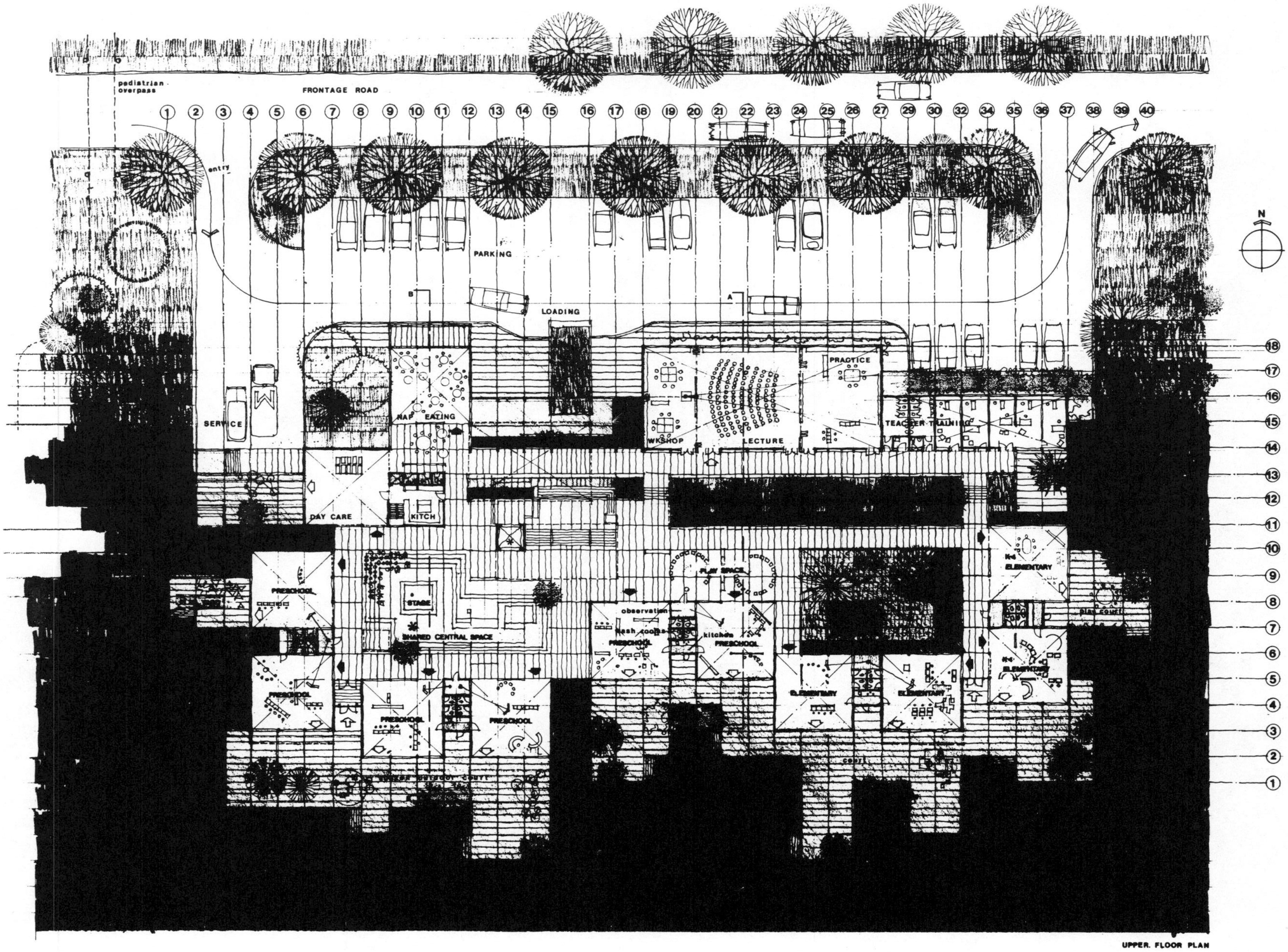

MONTESSORI SCHOOL & ADMINISTRATIVE BUILDING, MONTESSORI FOUNDATION OF MINNESOTA INC, ST PAUL MINNESOTA

UPPER FLOOR PLAN

SCALE : 1/16" = 1'-0"    GRAPHIC :

N

Pedestrian overpass

FRONTAGE ROAD

PARKING

UTILITY

SERVICE

LOADING

PRACTICE

TEACHER TRAINING

LECTURE

WORKSHOP

MAT-MAKING

KITCH

DAY CARE

PRESCHOOL

PLAY AREA

PRESCHOOL

Observation

PRESCHOOL

PRESCHOOL

SHARED CENTRAL SPACE

STAIR

PRESCHOOL

A

B

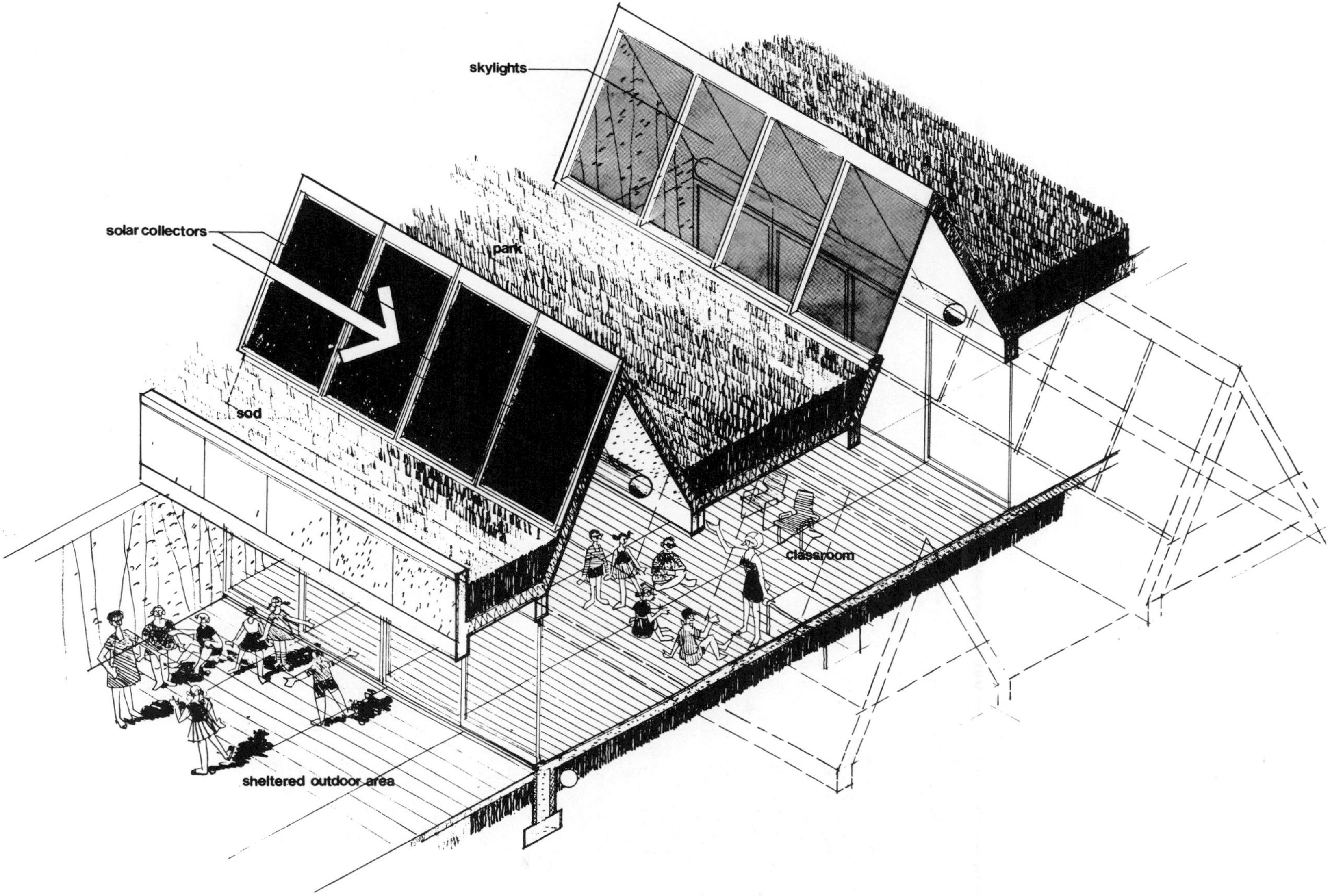

skylights

solar collectors

park

sod

classroom

sheltered outdoor area

SCALE : 1/16"= 1'-0"   GRAPHIC :

**MONTESSORI SCHOOL & ADMINISTRATIVE BUILDING, MONTESSORI FOUNDATION OF MINNESOTA INC, ST PAUL MINNESOTA**

## University Office Building for Reston
by Walter F. Roberts, Architect, Reston, Virginia

earth-sheltering moderates the yearly temperature fluctuations of the building as well.

The features incorporated in this building will reduce the estimated total energy costs to only one-third of those in a conventional building of comparable size.

One University Plaza, now under construction, began with the initial idea of an underground office, which would not only assist with energy conservation but would make the office structure less obtrusive within its residential area.

The 18,000 square foot building was integrated into a triangular shaped site with the exposed south elevation becoming the public face. On the east, the building cuts directly into the earth, while on the north and west, it is earth-bermed. Earth and rigid insulation cover the composite roof. This earth-covered roof creates a landscaped area, dotted by light monitors, which can be used as a recreation area for the office workers. A regional bike path also crosses this landscaped roof.

Notable energy-conserving features of the building include a south elevation with operable insulated glazing protected from the summer sun by a redwood sunscreen, as well as daylighting monitors rising above the earth-bermed structure itself. In the plan, the light monitors occur in the center of each square bay. The bays are grouped into six main spaces for the Housing Authority's six departments. Acoustical walls with integrated mechanical space form the divisions between these departments. The design of the monitors maximizes the use of daylight to reduce lighting costs 50% and provides a connection to the outdoors for each office. The monitors capture and reflect the natural light while photoelectric dimmers located within the monitors control the addition of artificial light when necessary to maintain a constant light level.

The light monitors also assist in the heating and cooling of the building by acting as exhaust chimneys or heat collectors. In the cooling process the building's 14 foot ceiling height assists in stratifying the hot air, which then rises into the monitors and is exhausted by automatically controlled louvers. During the heating process, solar radiation generates warm air within the monitors, which then collects at the ceiling allowing the warmth to be stored in the earth mass which returns the warmth at night. The

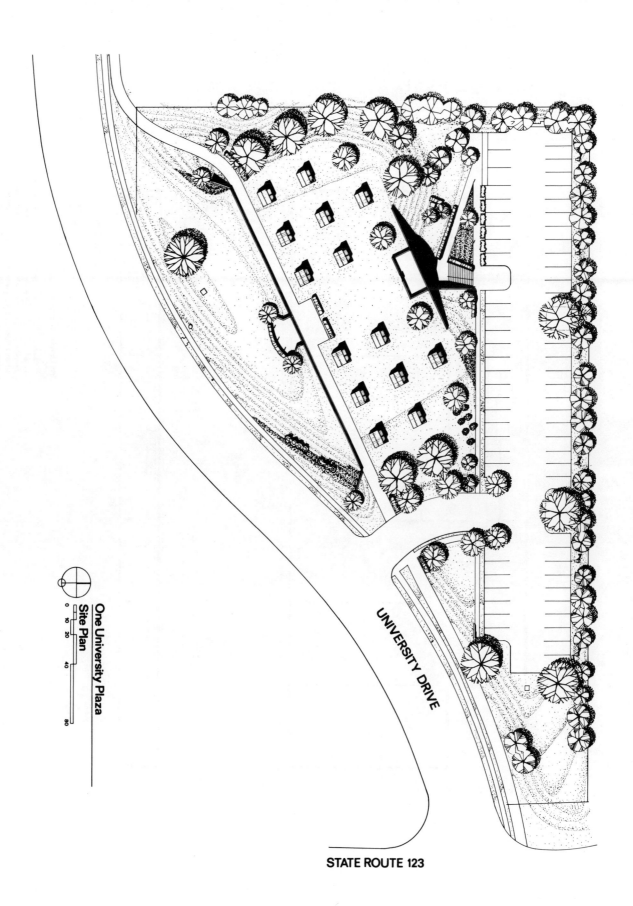

One University Plaza
**Site Plan**

0  10  20  40  80

UNIVERSITY DRIVE

**STATE ROUTE 123**

185

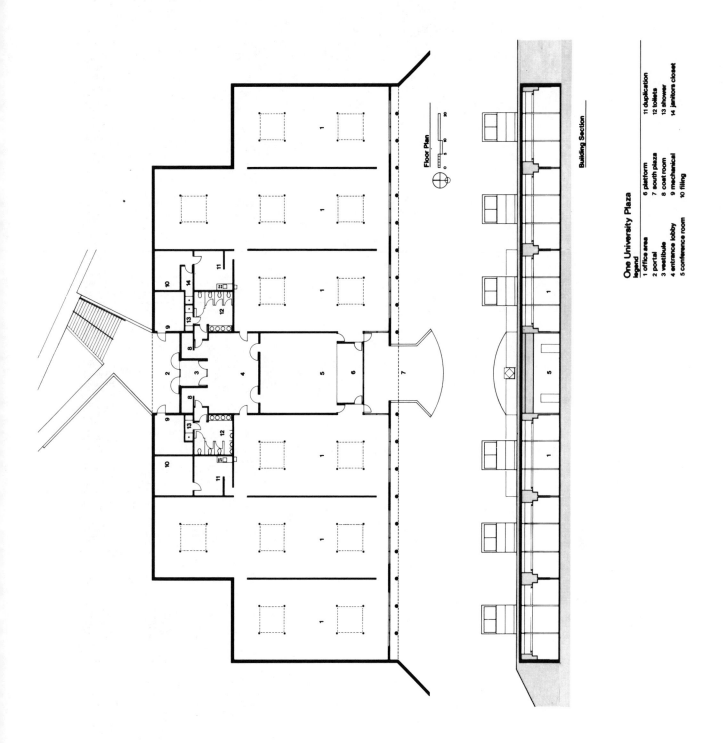

Floor Plan

Building Section

**One University Plaza**

legend

| | | |
|---|---|---|
| 1 office area | 6 platform | 11 duplication |
| 2 portal | 7 south plaza | 12 toilets |
| 3 vestibule | 8 coat room | 13 shower |
| 4 entrance lobby | 9 mechanical | 14 janitors closet |
| 5 conference room | 10 filing | |

# The George R. Moscone Convention Center in San Francisco, California by Hellmuth, Obata & Kassabaum, Inc., San Francisco, California

The George R. Moscone Convention Center will provide a total of 650,000 square feet of predominantly underground space for convention, meeting and exhibit use. Located on an 11 acre site in the Yerba Buena Redevelopment Area of San Francisco, an area that has been vacant since the 1960s, the center is expected to generate significant economic benefits for San Francisco's major industry, tourism, and to act as a catalyst for further development of the area.

The voters of the City of San Francisco adopted a proposition that established, as a statement of policy, that the convention center should be underground, if financially feasible. Community concern indicated a desire to keep the bulk of the convention center from becoming a barrier and to keep the rooftop available for cultural or recreational development. An added advantage of an earth-sheltered structure is that savings of approximately 25% of the total energy cost of heating and cooling the underground building will be realized when compared to a similar building above ground.

The roof of the convention center's exhibit hall is supported by eight pairs of post-tensioned concrete arches that span 275 feet. The arch frame concept gives the facility the distinction of being the only column-free exhibit hall of its size in the United States. The column-free design simplifies set-up and dismantling of large exhibits and allows the exhibit booths to be situated in a variety of configurations. The roof structure is capable of supporting the weight of three-story buildings or an eight-acre landscaped park with 50 foot high trees. Until further development occurs, the roof will be landscaped around the perimeter to create a park-like appearance.

The entrance lobby serves primarily as a registration area and is the only part of the convention center visible above ground. It is

surrounded on all four sides by glass that permits natural light to penetrate 30 feet below grade to the exhibit hall level. Windows between the exhibit hall and the lobby permit the flow of natural light into the hall and provide occupants of the hall with a view of the city's skyline. Thirty-one meeting rooms are located on the mezzanine level, where groups as small as 50 or as large as 600 can convene. Also on the mezzanine level are administrative offices and support spaces.

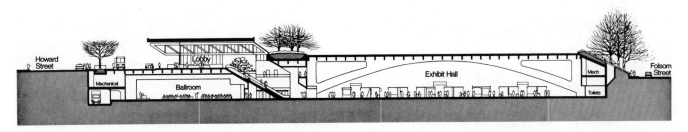

**Moscone Center
Transverse Section**

Hellmuth, Obata, Kassabaum, Inc. Architects
Young and Associates, AIA Associate Architects
T.Y. Lin International Structural Engineers
Hayakawa Associates Mechanical Engineers
SWA Group Landscape Architects
Turner Construction Co. Construction Manager

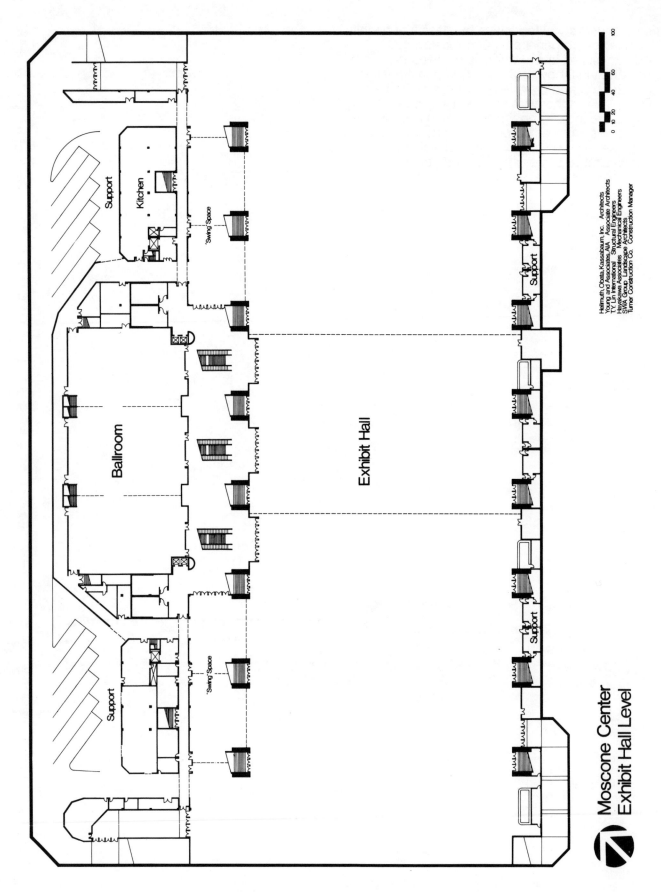

Moscone Center
Exhibit Hall Level

Helmuth, Obata, Kassabaum, Inc. Architects
Young and Associates, AIA Associate Architects
T.Y. Lin International Structural Engineers
Hayakawa Associates Mechanical Engineers
SWA Group, Landscape Architects
Turner Construction Co. Construction Manager

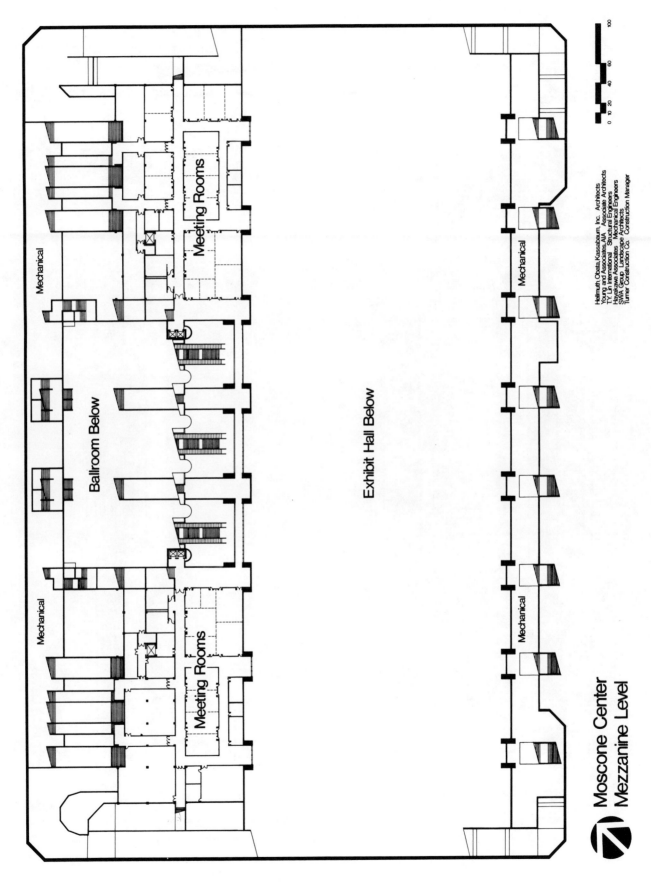

Meeting Rooms

Mechanical

Ballroom Below

Meeting Rooms

Mechanical

Exhibit Hall Below

Mechanical

Mechanical

0 10 20    40    60    100

Hellmuth Obata Kassabaum, Inc.  Architects
Young and Associates, AIA  Associate Architects
T.Y. Lin International  Structural Engineers
Hayakawa Associates  Mechanical Engineers
SWA Group  Landscape Architects
Turner Construction Co.  Construction Manager

## Moscone Center
## Mezzanine Level

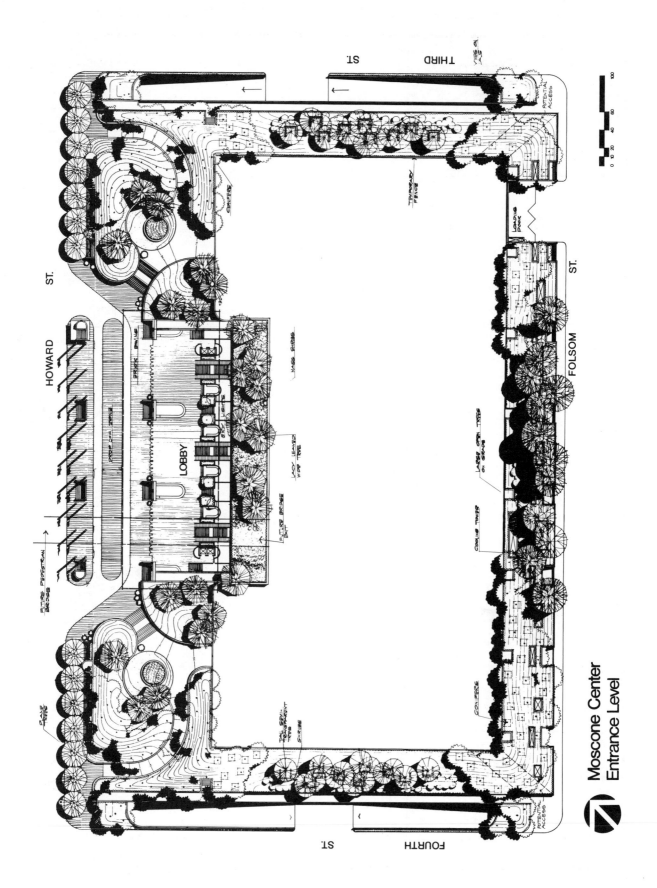

Moscone Center
Entrance Level

194

# First Award

## Master Plan for the Institute of Technology, University of Minnesota
by Brian R. Johnson, Student, University of Minnesota, Anoka, Minnesota

This Masterplan is a proposal for the long-range development of the Institute of Technology (I.T.) at the University of Minnesota. It proposes construction of facilities in three phases over 15 years. As well as providing much needed modern facilities to allow the school to maintain a national stature, the design creates a visual identity for I.T., integrating it with existing university systems.

Underlying the I.T. Masterplan are several major concepts that provide a general theme. These concepts tie diverse functions to each other as well as to the University as a whole.

Spatial organization is designed to preserve and create open space on campus. Large use of underground construction will allow the open spaces to be preserved and will create a series of enclosures and open spaces throughout the complex.

The main energy source for the complex will be the University of Minnesota district heat system but the load will be reduced greatly by passive energy methods. The building massing has a linear quality with south and west exposure. This will allow maximum lighting with minimum glazing. By utilizing natural light, the need for artificial light, the largest energy consumer, is reduced.

Solar heat is captured in interior spaces or through collectors mounted on the exterior for heating purposes. Major interconnections of the building massing allows the surface exposed to the weather to be reduced, thus cutting heat loss. Extensive underground development benefits from the earth's constant 55°F temperature significantly reduce heating and cooling requirements throughout the year.

Two buildings were developed within the Masterplan to give specific examples of the previously mentioned concepts. The first of these is the I.T. Center. The 86,000 square foot building will serve as a focus for I.T. It will contain the library, a lounge with display, eating and meeting functions and administration spaces. The second building developed will provide a north entrance to the I.T. campus. The 122,000 square foot Civil and Mineral Engineering building will contain labs, classrooms and a commons area.

These buildings are 90% underground and use passive solar heat gain techniques through the use of large south facing louvered skylights. Within these skylights are central atriums of masonry that act as heat storage and distribution points to serve the rest of the building.

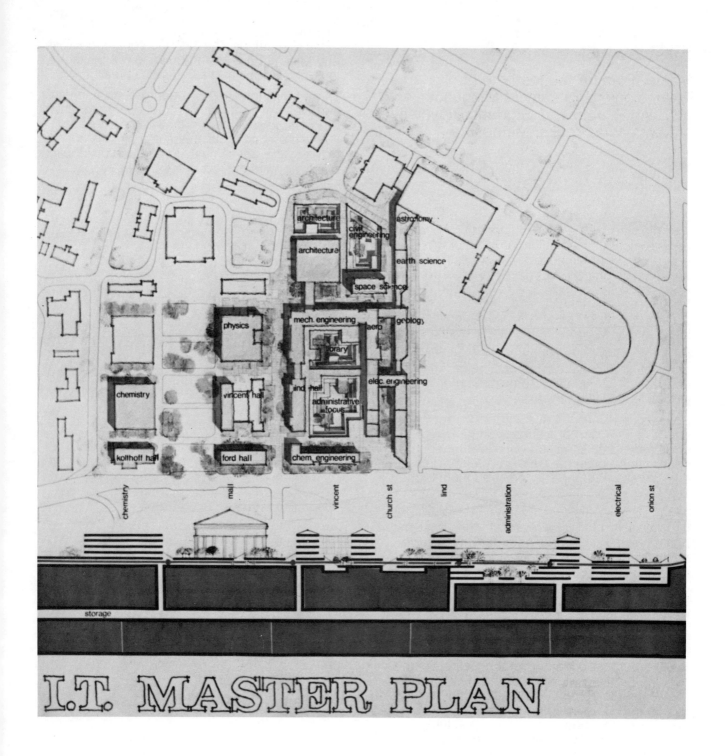

I.T. MASTER PLAN

I.T.
CENTER

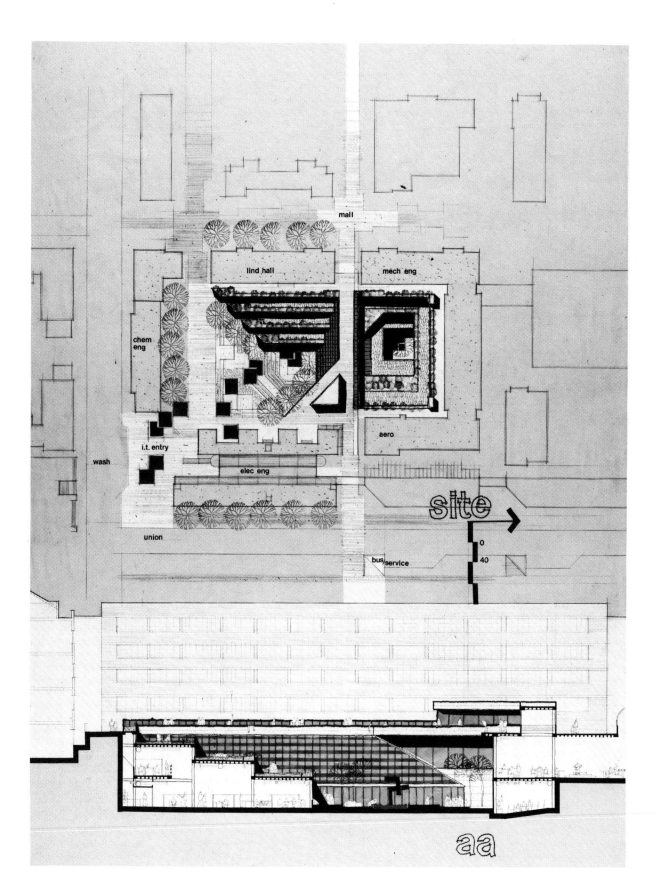

mall

lind hall

mech eng

chem eng

aero

wash

i.t. entry

elec eng

site

0

40

union

bus service

aa

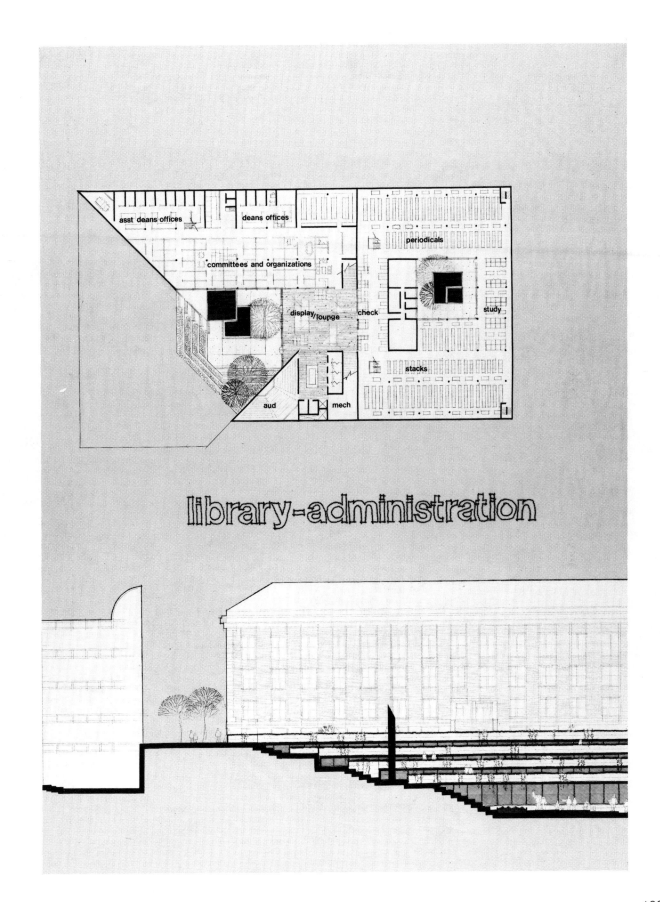

asst deans offices

deans offices

periodicals

committees and organizations

display/lounge  check

study

stacks

aud  mech

library-administration

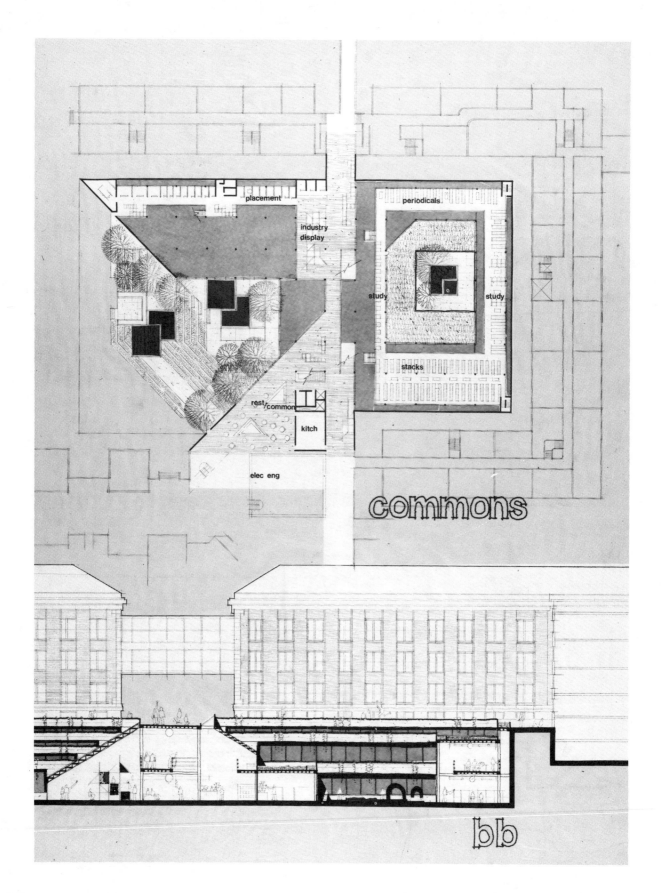

placement

industry
display

periodicals

study                study

stacks

rest/common

kitch

elec eng

commons

bb

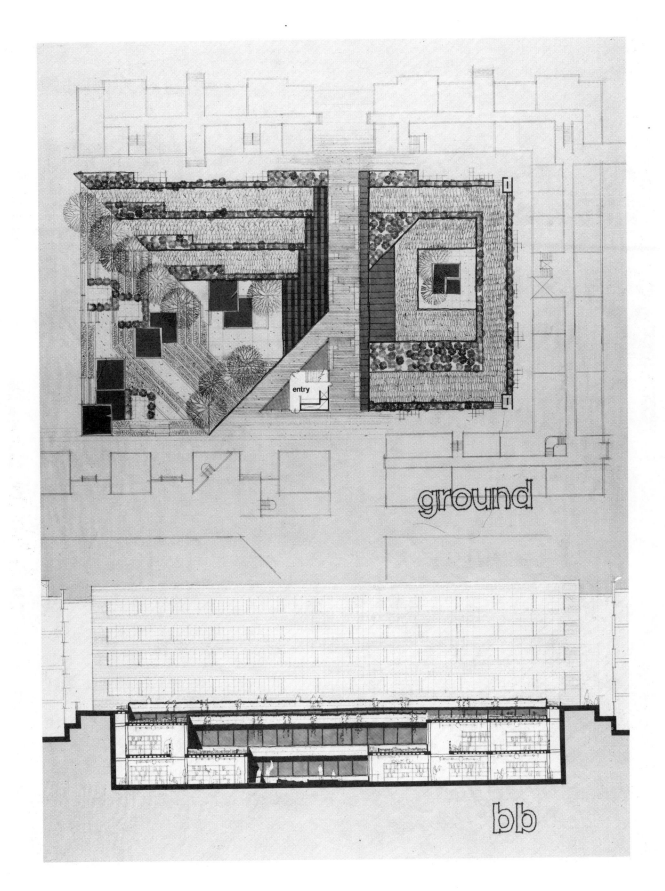

entry

ground

bb

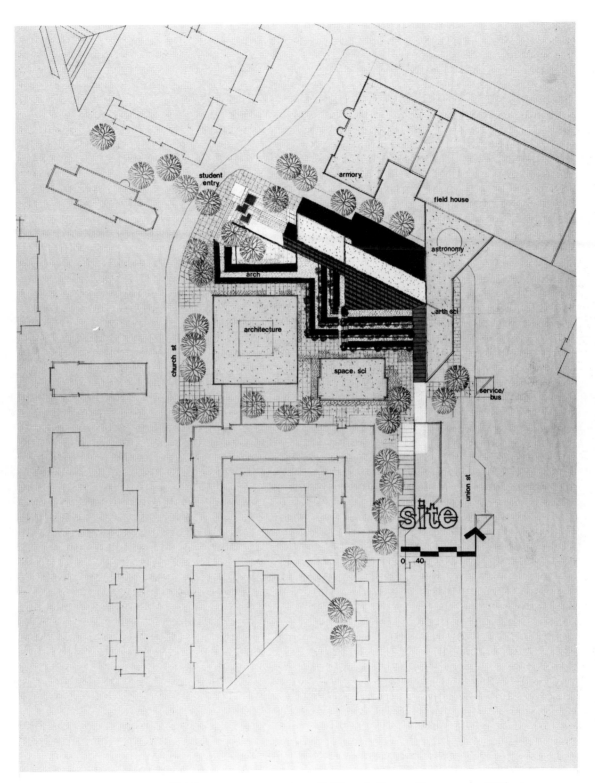

student entry

armory

field house

astronomy

arch

earth sci

architecture

space sci

service/bus

church st

union st

site

0 40

# CIVIL & MINERAL ENGINEERING

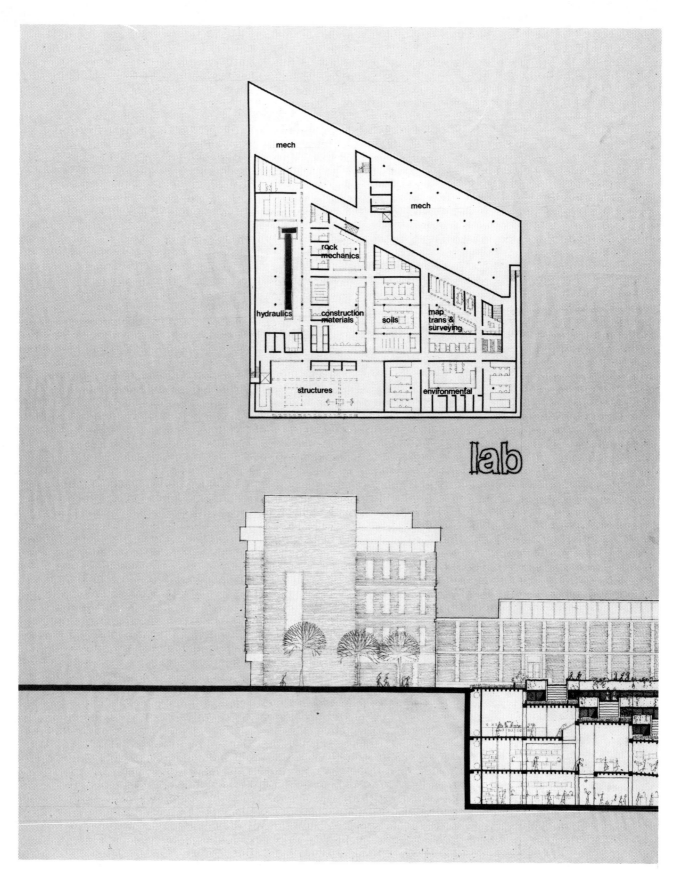

lab

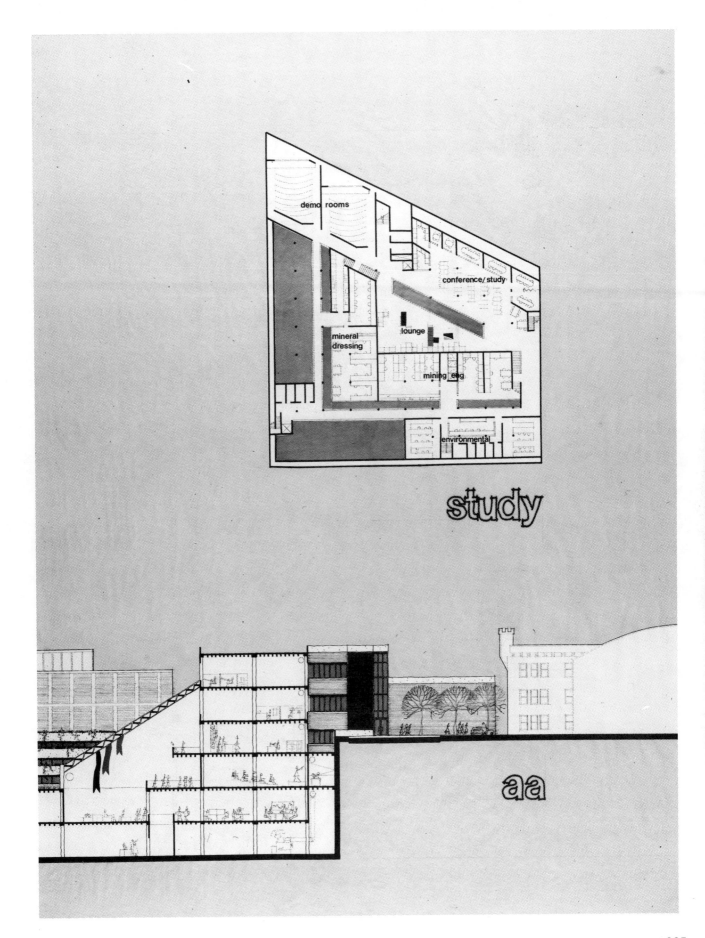

study

aa

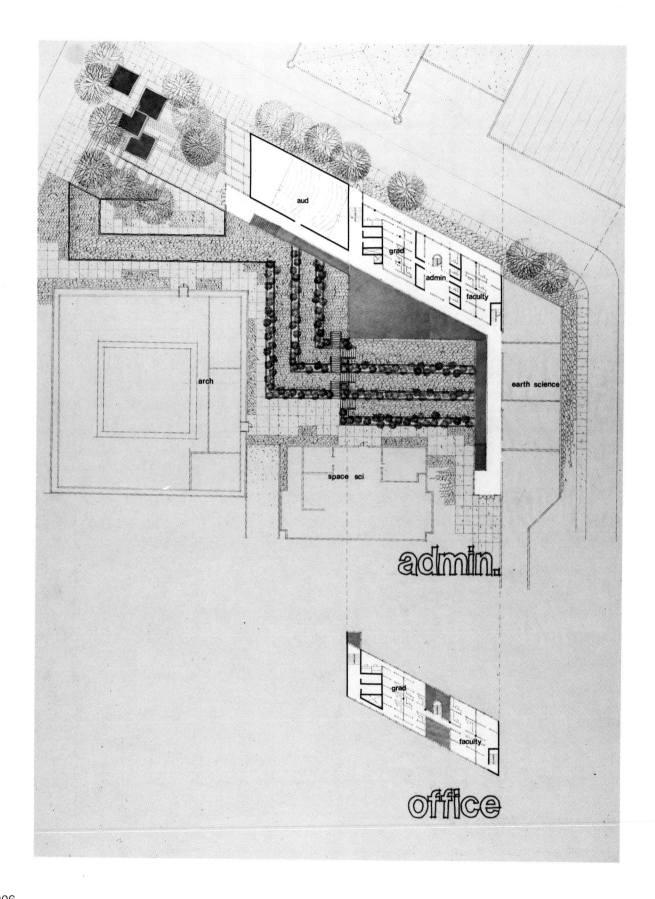

aud

grad

admin

faculty

arch

earth science

space sci

admin.

grad

faculty

office

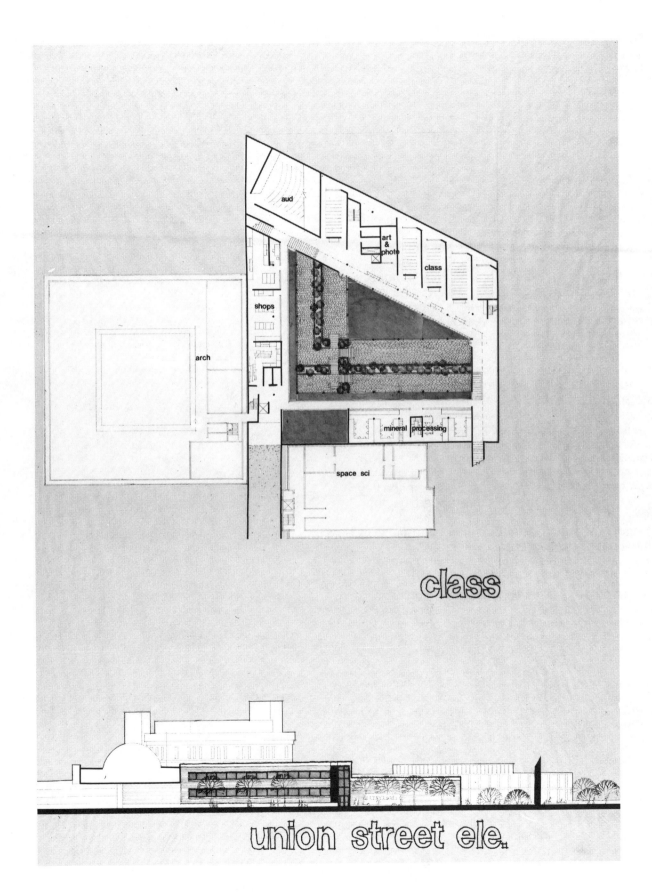

class

union street ele.

## Market Street Transfer Facility in Downtown Denver, Colorado by Cornelius R. DuBois, Associate, Johnson-Hopson and Partners, Denver, Colorado

insulation for the terminal roof. Direct solar gain occurs only at the two glass enclosures which house the stairs and escalators connecting the underground concourse to the plaza above. This central concourse will provide the only conditioned space in the structure. Heating and cooling are required in only 15,000 square feet of the 53,000 square foot building.

The Market Street Transfer Facility will provide the interface between the regional/express bus systems and the shuttle vehicles serving the new 16th Street Mall through the heart of Denver's central business district. Located in the unique historic district of lower downtown, now the focus of widespread rehabilitation efforts, the design of the transfer facility requires a sensitivity to urban issues far beyond the consideration of the operational program.

The placement of the express bus terminal functions below the grade of Market Street making possible the development of a city park extending to all corners of the block and providing a significant counterpoint to the Civic Center open space at the opposite end of the 16th Street Mall. A plaza development of this size accents the end of the Mall with a significant event and offers a sizable urban open space in an area that presently has little. It is hoped that the park will also provide an element of stability in a district now undergoing a period of transition, and that its presence will encourage further development oriented to a variety of human activities in the neighborhood.

The underground terminal affords operational advantages for the $7.5 million project as well as the ability to expand the program from seven to ten berths, a capability which was not available in an earlier on-grade scheme. The terminal configuration made possible by the below grade format allows routing buses around the perimeter of a central pedestrian concourse. Containing bus fumes and noise beneath the street level will aid in the integration of the Market Street Transfer Facility into a quiet neighborhood. This plan not only offers optimum efficiency but also creates a secure, easily-managed space, a critical concern for a facility which will accommodate over 10,000 patrons a day.

Energy considerations were also an impetus for the underground placement of the facility. The plaza construction, four to five feet of foam blocks and earth fill for planting, will offer substantial

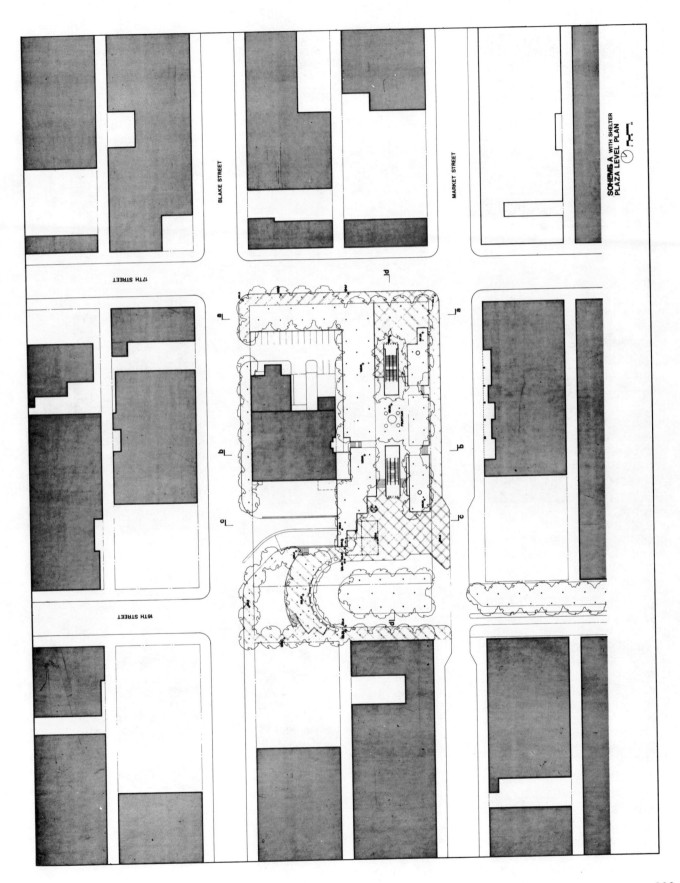

BLAKE STREET

MARKET STREET

17TH STREET

16TH STREET

SCHEME A WITH SHELTER
PLAZA LEVEL PLAN

209

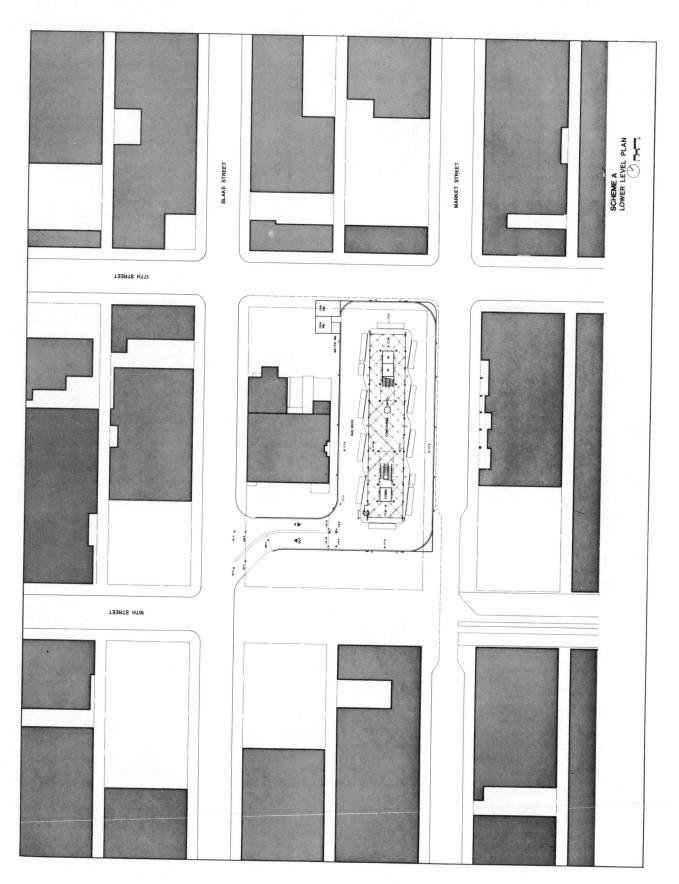

BLAKE STREET

MARKET STREET

17TH STREET

16TH STREET

SCHEME A
LOWER LEVEL PLAN

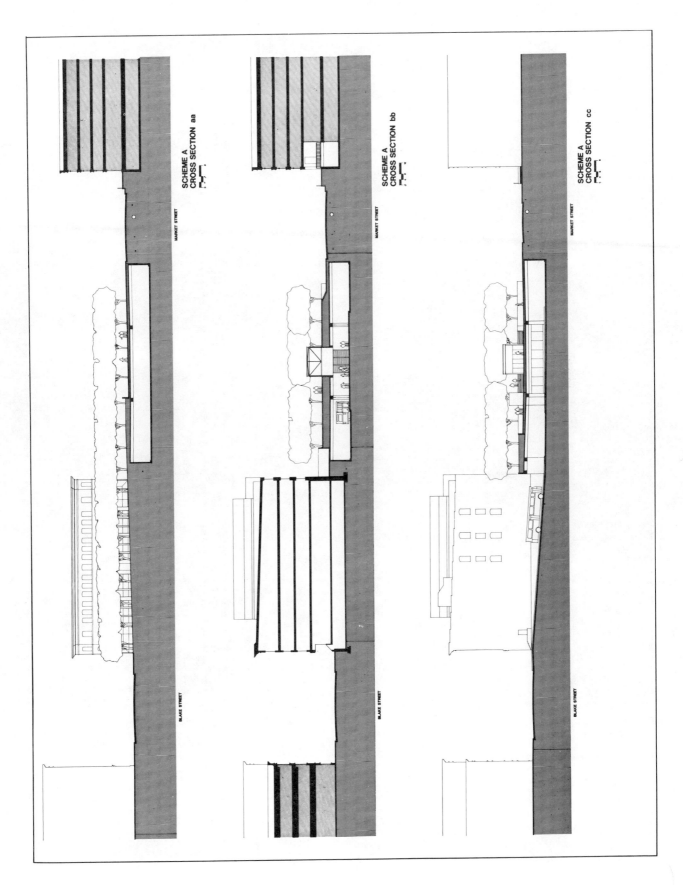

SCHEME A
CROSS SECTION aa

SCHEME A
CROSS SECTION bb

SCHEME A
CROSS SECTION cc

MARKET STREET

MARKET STREET

MARKET STREET

BLAKE STREET

BLAKE STREET

BLAKE STREET

211

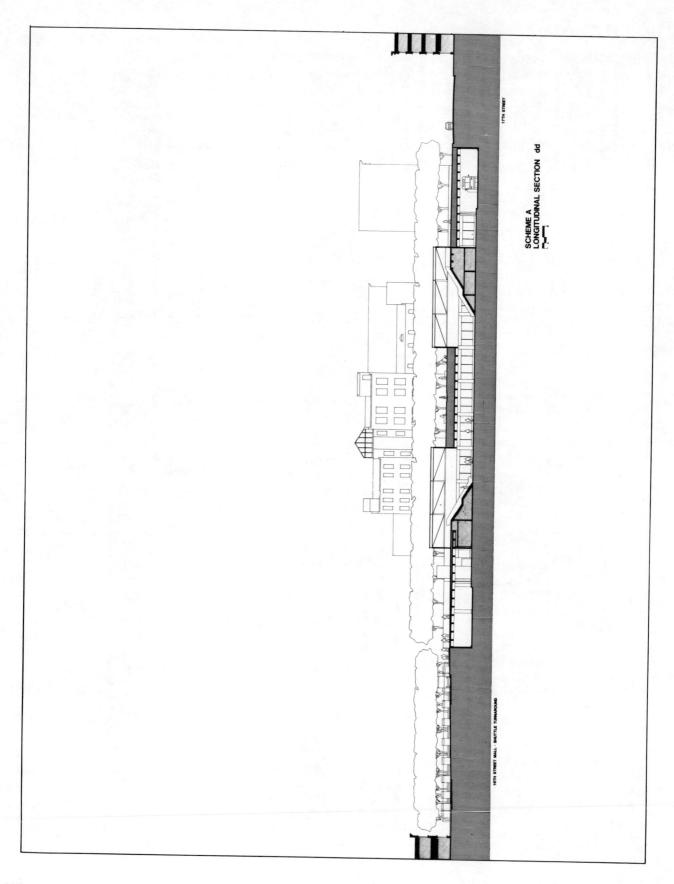

SCHEME A
LONGITUDINAL SECTION dd

17TH STREET

16TH STREET MALL - SHUTTLE TURNAROUND

212

# Merit Award

## A Fallout Shelter Recycled as a Library for the Physically Handicapped in Watertown, Massachusetts
by John Coughlan, Arrow Street, Inc., Cambridge, Massachusetts

The regional library serves more than 9,000 blind and physically handicapped patrons in Massachusetts and New England. The library serves its clientele principally by providing them by mail with talking books (discs and cassettes) and braille volumes.

These heavy and bulky materials are presently stored in very inadequate space in the school's own library. This space, approximately 6500 square feet, is not only too small but is spread over three floors, a serious liability considering the weight and size of the materials. Further, once collected for dispatch to readers, the materials have to be sent from a small and inconveniently located shipping and receiving area.

An outmoded and neglected underground fallout shelter of poured concrete construction is located on the campus of the Perkins School. The School's administration commissioned a design study to see if this structure and its space could be adapted to serve the regional library.

The design study demonstrated that the shelter could be revitalized to serve as a regional library facility that was large enough for adequate storage, convenient for shipping and receiving, barrier-free for patrons who might want to visit the library themselves and a congenial workplace for staff and library users.

The main floor of the shelter will accommodate, on a single floor, a storage area nearly three times as large as the existing stack areas. Conveniently adjacent to these stacks and on the same floor are a greatly enlarged shipping and receiving area, repair facilities for mechanical equipment used by clients, other work areas for staff, a lobby and conference room and a public reading room. An upper level has been created from an existing passage between the outside entrance to the shelter and the stairs leading to the main level. This upper level will serve the library as a special mezzanine reading area.

The southeastern edge of the shelter, now extending out of the hillside into the nearby Charles River valley, has been opened further to provide a large shipping dock and an entry point with adjacent parking for patrons.

Four skylights are located on the roof of the revitalized structure to provide natural daylight and passive solar collection for the earth-sheltered, south-facing structure. The architects have planned a scent garden to be planted under these skylights as one of the many special amenities designed for the library's blind and handicapped patrons.

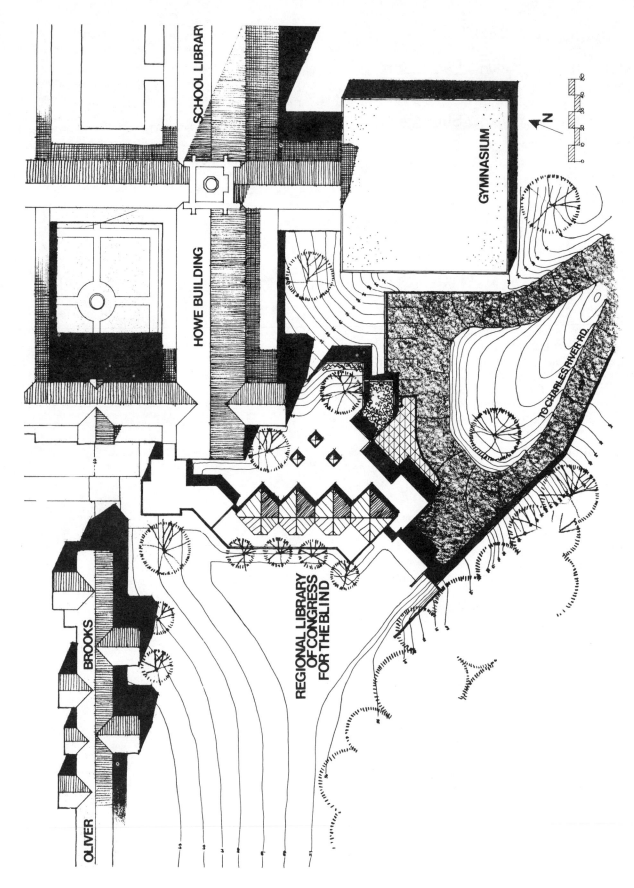

SCHOOL LIBRARY

GYMNASIUM

N

HOWE BUILDING

TO CHARLES RIVER RD.

REGIONAL LIBRARY
OF CONGRESS
FOR THE BLIND

BROOKS

OLIVER

214

SECTION A - A

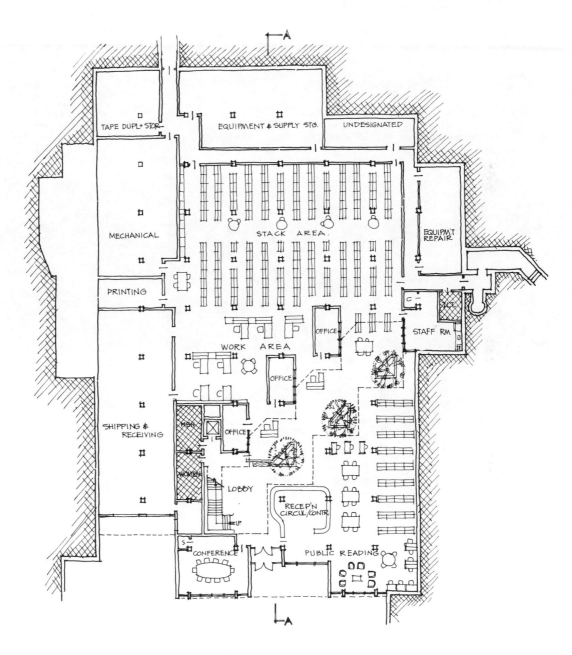

PLAN - MAIN LEVEL

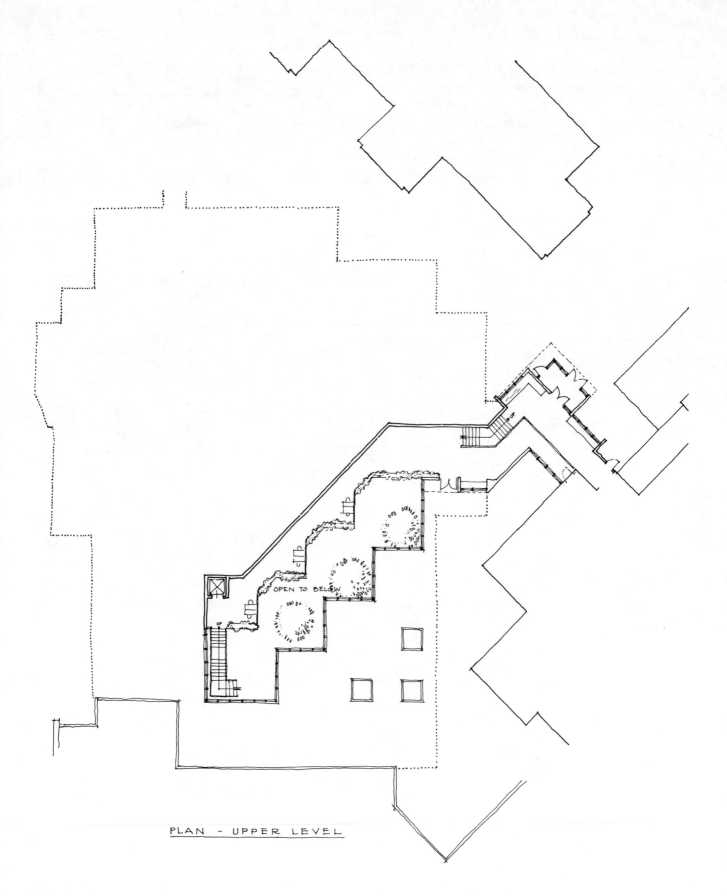

PLAN - UPPER LEVEL

# Merit Award

## Iowa State Historical Museum near the Capitol Complex
by Herbert M. Stone, Brown Healey Bock, Cedar Rapids, Iowa

The most important design premise was to provide a safe and secure location for the State's vast historical collection. Many sites on the Capitol grounds were studied for their suitability as a location for a new museum. The site that was eventually selected is south of the existing State Capitol in an under-used, "park-like" area that was poorly maintained and insufficiently used. There is an existing bridge at this location which provides access from the Capitol and has to be replaced due to its deterioration. By combining a new building with the new pedestrian bridge, a safe pedestrian link to the Capitol grounds itself would be provided. The integration of dual site development by the connection of the building to the bridge extends the area of the Capitol grounds to this location and enhances the total surroundings. The development of an underground structure at this site eliminates a potential mass that might distract from the Capitol and creates significant outdoor display spaces for larger museum pieces, such as trains, farm equipment, etc., which are to be exhibited. The design allowed the natural qualities of the site to be retained and enhanced. It also allows pedestrian and vehicular circulation to be developed at different levels without conflict.

The essential feature of the museum program is to provide ample area for the display of varying kinds of exhibits requiring different amounts of space and great areas of unencumbered walls. A building of conventional construction on a more restrictive site would have required a massive and overpowering structure. A solution of this type allows a blend of energy efficient design, good security, natural acoustical quality of the surrounding earth mass and the utilization of a natural controlled daylight. It is practical to meet a certain percentage of the building's heating and cooling needs utilizing solar energy. The collectors will be placed on a series of earth berms south of the main building down the slope.

The site design has been projected as a series of ground level terraces allowing for the development of the area with one major architectural feature — a large, jewel-like skylight, which allows controlled natural light to filter through into the lower levels of the building. The exterior above grade portion of the building and site has been designed to reflect the classical ground pattern relationships and formal massing of the Capitol Complex. The created interior volume becomes the most important aspect of the architectural solution providing a satisfactory stage for the display of various museum exhibits. The building has been stepped in section in order to avoid a large rock outcropping which encroaches into the north boundary of the site.

The main vehicular approach to the building is underneath the bridge, which creates a contemporary "porte-cochere." The entrance is located off center in order to protect it from the prevailing winter winds, which come from the northwest and southwest in this region. The pedestrian approach from the Capitol is designed in such a manner as to draw people across the terrace area, with a glimpse into the interior from above. At the south end of the upper terrace, a series of steps flow around the main area down to a sheltered terrace on the south side of the complex, where existing site features have been retained. A small Japanese bell pavilion is the only site feature at the level of the new upper terrace on the axis of the Capitol. It has been placed in an honored position in relation to the new museum, where it is accessible from the exterior as well as visible from the interior.

The main exposure of the building is to the east, which allows the external mass of the building to be downplayed. The surrounding natural grades of the area allow very little of the building to be perceived from a distance in any direction, which was a requirement in allowing for the use of this site.

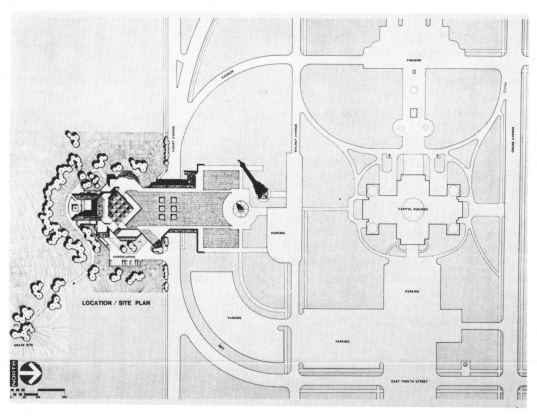

LOCATION / SITE PLAN

GRAVE SITE

NORTH

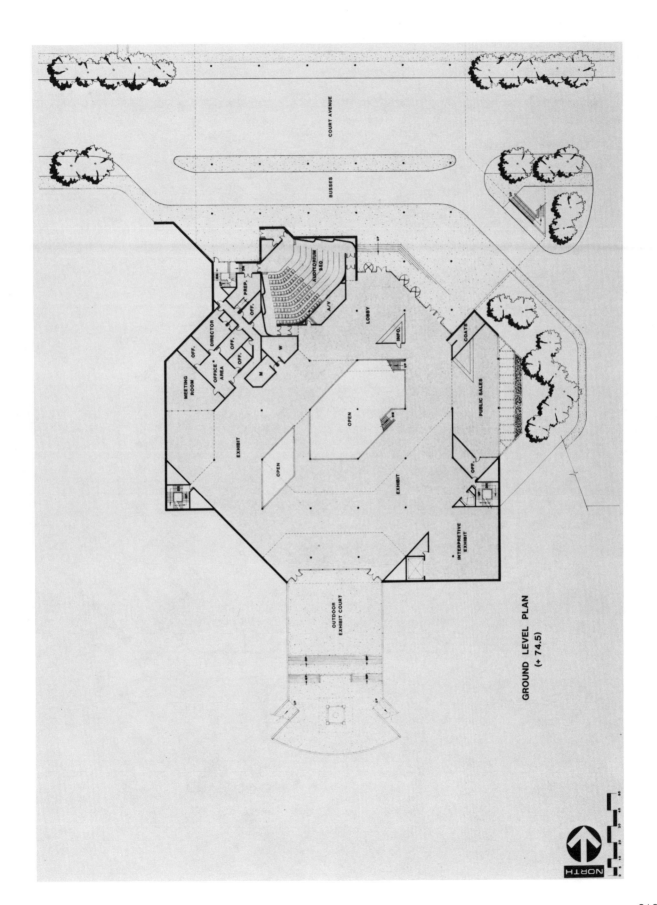

GROUND LEVEL PLAN
(+ 74.5)

COURT AVENUE

BUSSES

AUDITORIUM
250

PREP.

OFF.

A/V

DIRECTOR

OFFICE
AREA

OFF.

OFF.

LOBBY

INFO.

COATS

MEETING
ROOM

M

W

EXHIBIT

OPEN

OPEN

PUBLIC SALES

OFF.

EXHIBIT

INTERPRETIVE
EXHIBIT

OUTDOOR
EXHIBIT COURT

NORTH

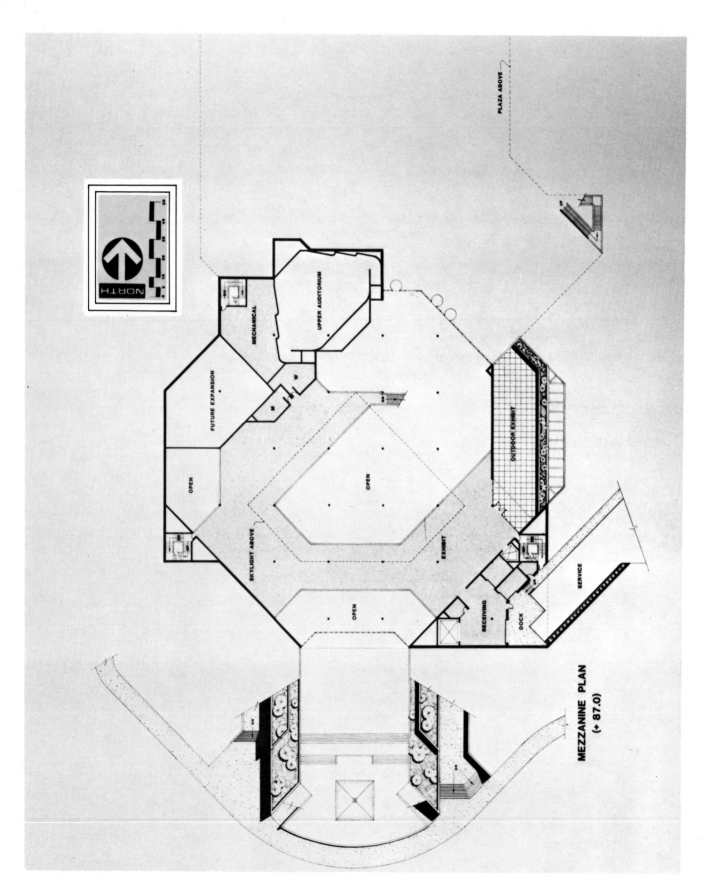

MEZZANINE PLAN
(+ 87.0)

NORTH

220

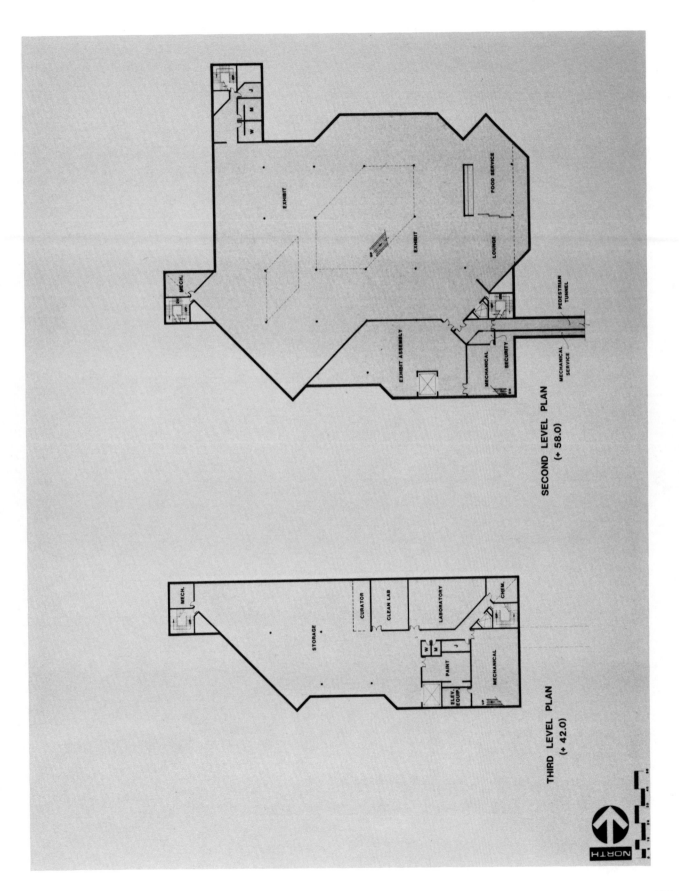

SECOND LEVEL PLAN
(+ 58.0)

THIRD LEVEL PLAN
(+ 42.0)

NORTH

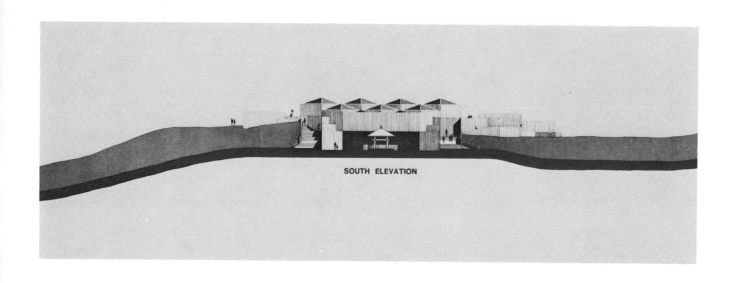

SOUTH ELEVATION

EAST ELEVATION

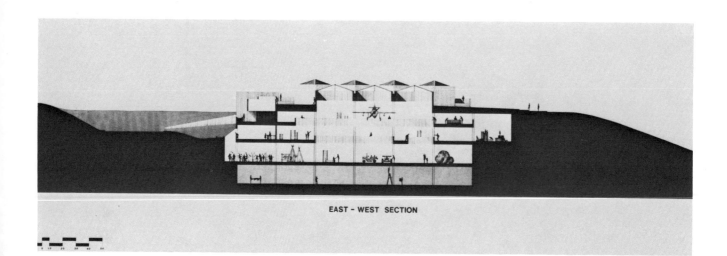

EAST - WEST SECTION

## Library Addition for a Midwestern University
by Walter Scholer, Walter Scholer and Associates, Inc., Lafayette, Indiana

This project consists of a 100,000 square foot library addition for a major midwestern university. The facility will house 1,000,000 volumes and provides study space for approximately 1000 students.

The program committee felt very strongly that the project should be an addition to the existing facility rather than a separate building. An enclosed connection between this facility and the existing facility was also required. The most logical site for a conventional building addition was to the north of the present facility, a space occupied by a parking lot. However, this site proved to be impractical since the cost to relocate the existing tunnels, utility pipes, and electrical conduits serving the campus was prohibitive. As a result, it was decided to locate the new facility directly in front of the main facade of the present facility.

In order to maintain the present building facade and to preserve the aesthetics of this area of the campus, the decision was made to design and build an underground structure.

The design consists of a two-story underground library with a landscaped terrace on top. The terrace has seating areas for student gatherings and ramped walks for easy access for the handicapped and small emergency vehicles.

Lightwells along the south side of the structure allow natural light to penetrate into the student study areas. These light wells are heavily landscaped and slope up to finish grade at street level.

The structure consists of a reinforced concrete flat-slab floor and roof system supported on concrete columns. Columns have been spaced on a 26 foot square bay in order to accommodate the electrically operated compact stack units. Perimeter walls are poured-in-place concrete. The roof deck is sloped and covered with protective waterproofed membrane assembly. Four to five feet of earth will be placed on the roof deck providing protection for the waterproofing

membrane, ample soil for landscaping and a moderating envelope. Cantilevered piling is provided to protect the foundation system of the existing facility.

Heating and cooling loads at the perimeter of the building are reduced due to the constant temperature of the earth envelope. It has been projected that the heating cost due to heat loss will be 15% less than a comparable above ground structure.

A specially designed floating ceiling system has been incorporated in order to give the owner the flexibility to change the direction of light fixtures easily as furnishing layouts may dictate. Light levels provide an 80 foot candle light level, resulting in a load of 1.43 watts per square foot. All diffusers and automatic fire extinguishing systems have been integrated into the floating ceiling system.

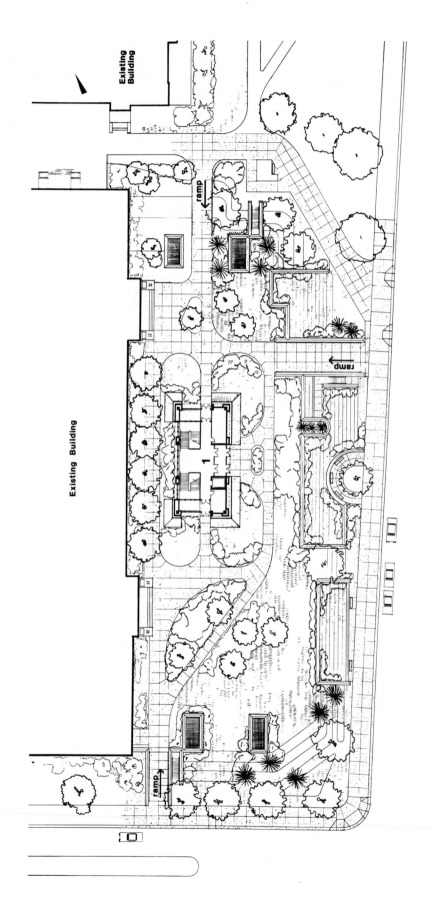

Existing Building

Existing Building

ramp

ramp

ramp

Site & Mezzanine Plan

0    20    40    60

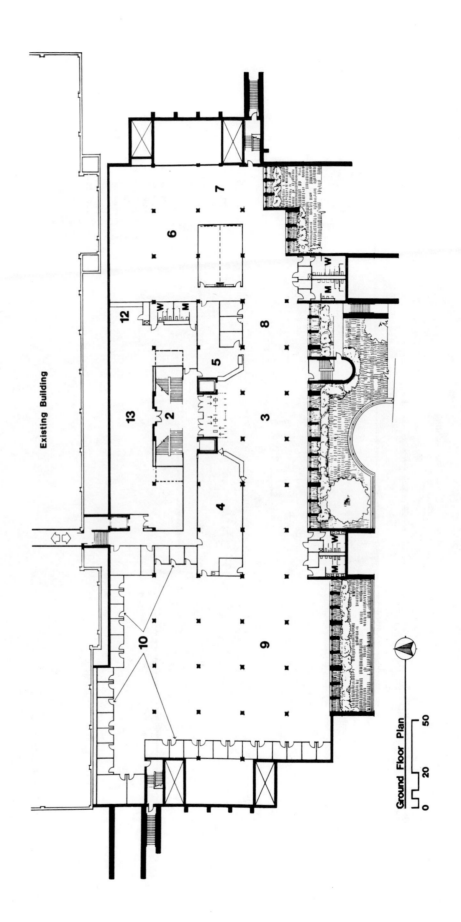

Existing Building

13

12

2

5

4

6

7

8

3

10

9

W
M

M
W

M
W

Ground Floor Plan

0    20    50

225

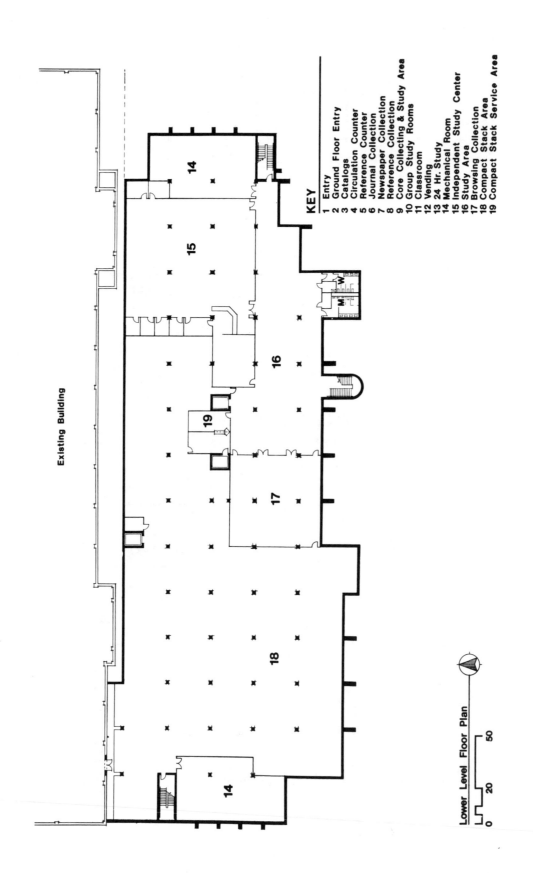

Existing Building

**KEY**

1 Entry
2 Ground Floor Entry
3 Catalogs
4 Circulation Counter
5 Reference Counter
6 Journal Collection
7 Newspaper Collection
8 Reference Collection
9 Core Collecting & Study Area
10 Group Study Rooms
11 Classroom
12 Vending
13 24 Hr. Study
14 Mechanical Room
15 Independent Study Center
16 Study Area
17 Browsing Collection
18 Compact Stack Area
19 Compact Stack Service Area

Lower Level Floor Plan

0        20        50

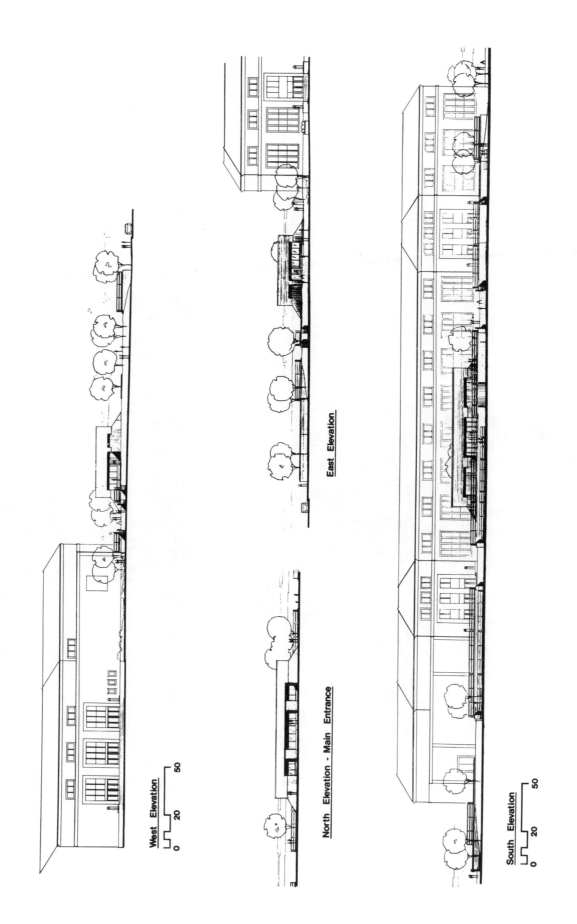

West Elevation

North Elevation - Main Entrance

East Elevation

South Elevation

227

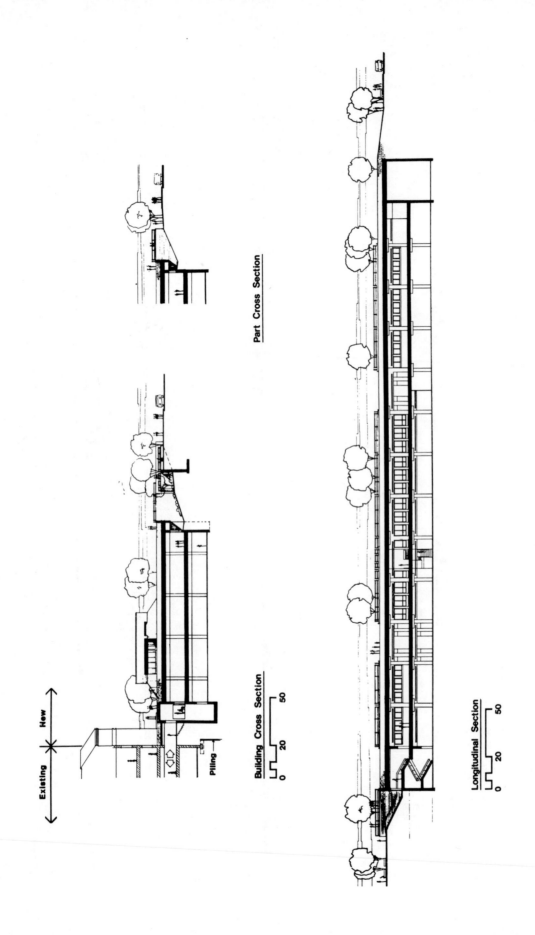

Part Cross Section

Building Cross Section

Longitudinal Section

Existing    New

Piling

0    20    50

0    20    50

228

# Honorable Mention

## Extension of the School of Architecture at Oklahoma State University
by Richard Dunham, Student, Oklahoma State University, Stillwater, Oklahoma

The goal is to design an underground extension to the School of Architecture directly south of the school of architecture. The site's southern exposure is unobstructed by a bordering facility.

The structure is designed with open courts on either side of the seminar space to facilitate passive solar gain. A half barrel vault protects the structure from heat build-up during summer months. This barrel is designed to help the structure gain heat during the winter months by allowing light to strike the brick floor. The brick will act as a heat sink which will help heat the research space during the peak heating periods. The structure will be underground to further

maintain constant temperatures. With the passive solar and underground design, the project is intended to facilitate underground building research and underground extension seminars.

The structure consists of poured-in-place columns and bearing walls in conjunction with a pan joist ceiling system to support the earth above.

Daylighting is used employing the barrel vault in conjunction with lightwells around the seminar space and above the office spaces on the north side of the building. In addition, heat is exhausted through the ventilation device in the barrel vault. The chimney effect in this vault area reduces the heat build-up in the critical summer months.

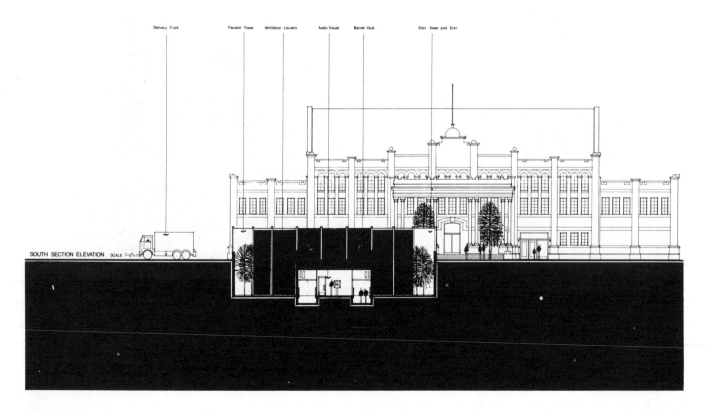

SOUTH SECTION ELEVATION    SCALE 1'-0" = 1/8"

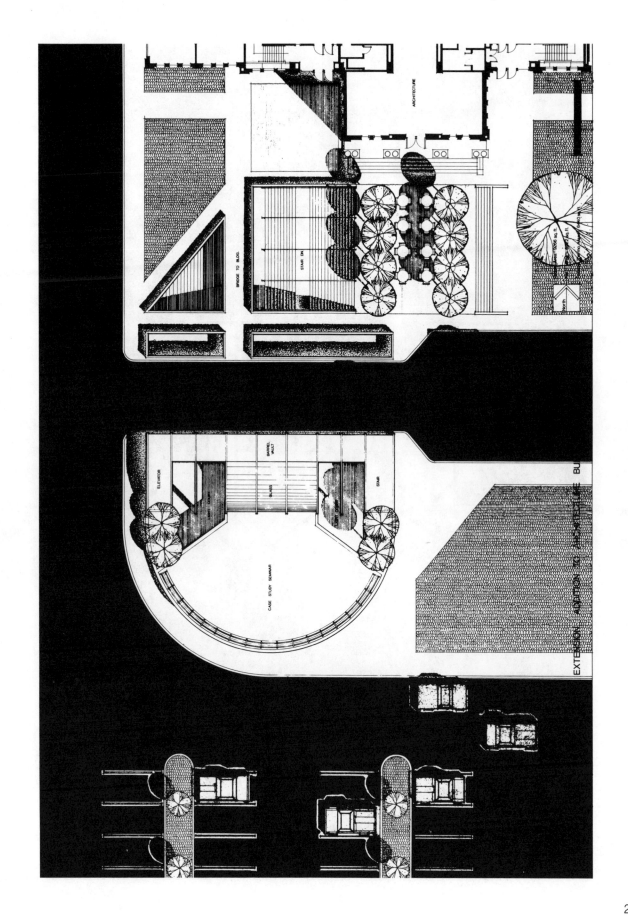

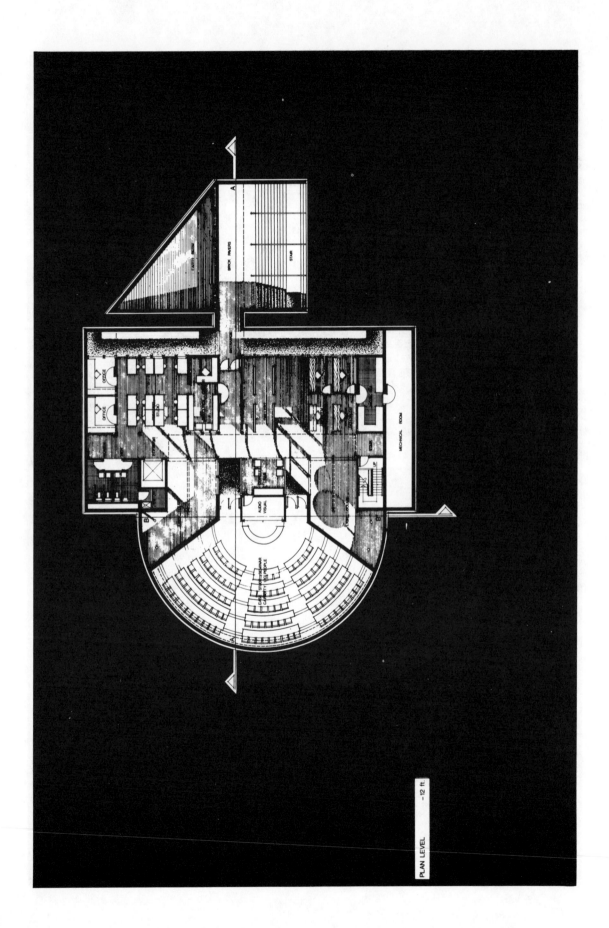

PLAN LEVEL  –12 ft.

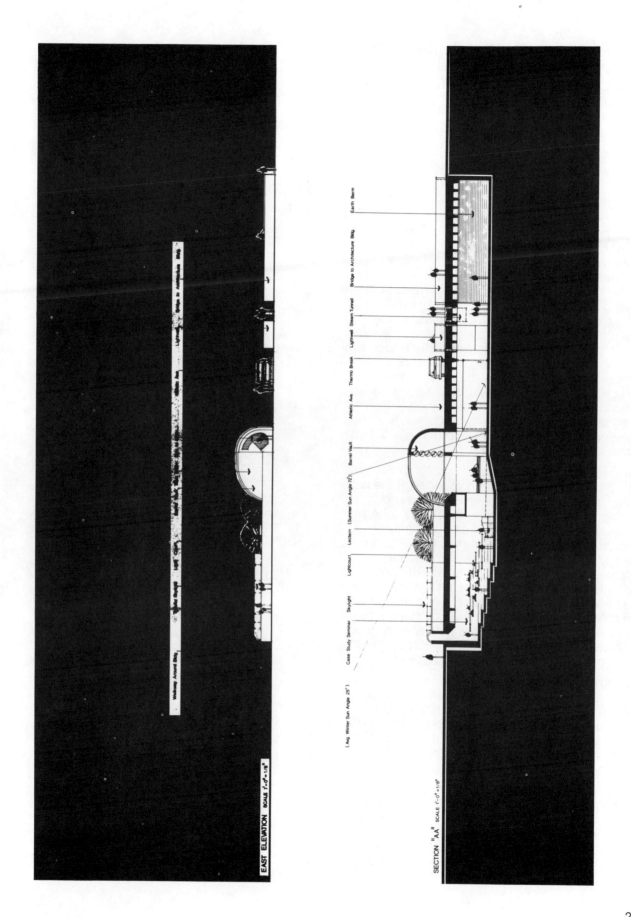

EAST ELEVATION SCALE 1'-0" = 1/8"

Walkway Around Bldg.

Lightwell, Bridge to Architecture Bldg.

SECTION "AA" SCALE 1'-0" = 1/8"

( Avg Winter Sun Angle 25° )

Case Study Seminar   Skylight   Lightcourt   Lectern   (Summer Sun Angle 72°)   Barrel Vault   Athletic Ave.   Thermo Break   Lightwell   Steam Tunnel   Bridge to Architecture Bldg.   Earth Berm

# Italian-American Cultural Center Addition for New York City
by Secundino Fernandez, Architect, DAT Consultants, Ltd., New York, New York

A stepped garden provides a double level entry/exit with circumferential parking drive and ceremonial drop-off. The lower entry loggia provides lobby and exhibit space leading to the central atrium. This sunken atrium formed from the existing house's basement becomes central circulation to the spaces around it and to the house above by means of a new stairwell leading the visitor to the promenade deck. The upper level promenade contains parking and existing monuments and relics with the addition of a memorial garden surrounded by the colonnaded signoria built by donors honoring notable Italian-Americans. Retaining existing grade levels at the promenade enables preservation of many of the existing trees.

The unique slope of the site allows for the penetration of natural light on all sides of the new facility minimizing energy consumption and helping in the preservation of the neighborhood's character. Skylight strips run parallel on three sides of the house with the inner strip defining the main circulation and the outer strip lighting perimeter exhibit and reading areas. The outer skylights are protected from heat exchange by trees and planters which also buffer parking from perimeter sidewalks. The covered loggia on the fourth side of the house allows natural light to penetrate the new facility while providing a view of the stepped sculptured garden outside.

The museum was designed as an underground structure both for energy conservation and for maximum protection of the existing landmark building from encroaching new construction. The clients and a group of professionals selected this design from 42 other submissions mainly for those two reasons.

The sponsoring organizations, The Order of the Sons of Italy and the New York State Association of Architects/American Institute of Architects, are now in the process of acquiring the necessary funding for the construction of the first Italian-American Cultural Center and Museum in the United States.

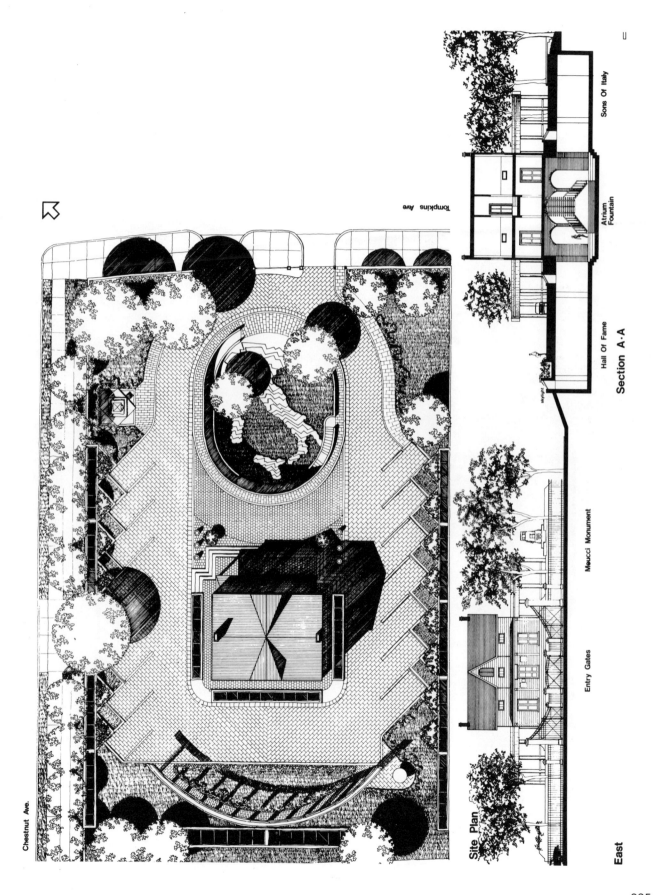

Chestnut Ave.

Tompkins Ave

Site Plan

Section A·A

Sons Of Italy

Atrium Fountain

Hall Of Fame

Meucci Monument

Entry Gates

East

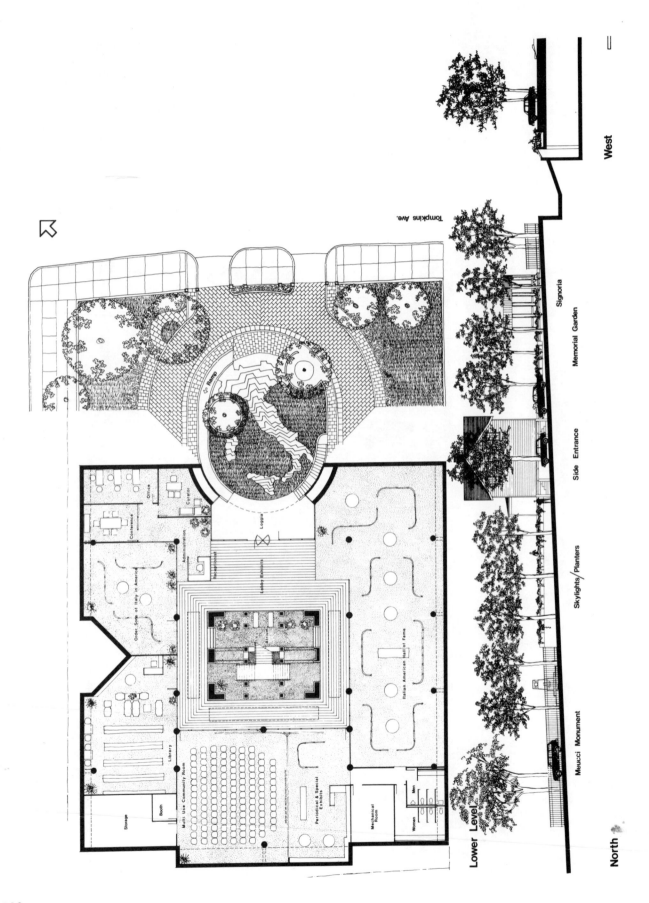

Lower Level

North

West

Tompkins Ave.

Signoria

Memorial Garden

Side Entrance

Skylights/Planters

Meucci Monument

Storage

Booth

Library

Multi-Use Community Room

Periodical & Special Exhibits

Mechanical Room

Men

Women

Office

Conference

Curator

Administration

Receptionist

Order: Sons of Italy in America

Lobby Exhibits

Loggia

Atrium

Italian American Hall of Fame

Ramp

236

**Section  B-B**

237

## Branch Municipal Center for Greenwich Village
### by Richard H. Clarke, Gwathmey Siegel Architects, New York, New York

The Branch Municipal Center was born out of a stated desire of the People of New York to improve the structure and operation of city government. The major premise of this improvement was that decentralization of municipal government should coincide as often as possible with existing communities and historical districts so that branch centers would best serve and, hopefully, reflect the interests of the particular community within its jurisdiction. Each center would serve a population of 100,000 to 150,000 with approximately 20,000 square feet of public service spaces including: offices for local representatives of the school district, welfare, police, fire and parks department, the planning board, etc., as well as a large public assembly hall, a smaller board meeting room, and community display space.

The Greenwich Village site is an island created by a shifted-grid intersection of Greenwich Avenue with Avenue of the Americas (6th Avenue) at West 9th Street. The island is dominated by the finest architectural product of the collaboration of Withers and Vaux, the ornate Jefferson Market, which recently became the local public library. The remainder of the triangular island has been made into a private garden which is cared for by members of the Village community.

The concept of the final scheme was to project an attitude of architectural conservation and maintenance for localized city government. The major idea was to maintain a community garden and horizontally layer the access of sunlight into the architecture in relation to the function of the spaces. The grid of the garden is oriented to Greenwich Avenue, representing the Municipal Center's relation to the local community as opposed to 6th Avenue which relates to Manhattan, and is composed of glass block pathways which light public circulation on the office level below. The objects on the garden level have a dual purpose; on the one hand, they are the signals of entry, vertical circulation, and public space, and on the other hand, they represent

garden elements and provide a low-scaled street edge and a point of observation over the whole neighborhood atop the off-axis entry pavilion.

Entry to the office level is made via any of the above grade steel and glass structures which also bring light into the concrete frame coffered slab structure. The office partitions are glass for greatest transference of natural ambient and artificial light. Mechanical systems run through the overhead coffers. The lower level contains the meeting and mechanical rooms which programmatically call for artificial light only. Both levels receive light from the three-story greenhouse which houses the escalator system, and skylights wash the walls and foundation of the Jefferson Market on the side opposite the atrium. Other amenities include an exterior pool on the office level which is shared by a cafe and a small reference library, and a pair of fountains along the arcade on the assembly level which provide both ample building humidity and serve to complement the quiet of an underground architecture.

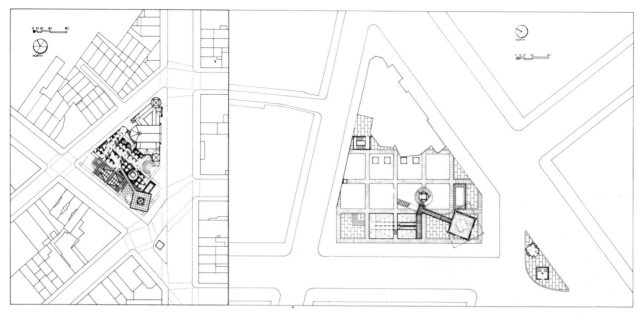

Site plan

Entrance level

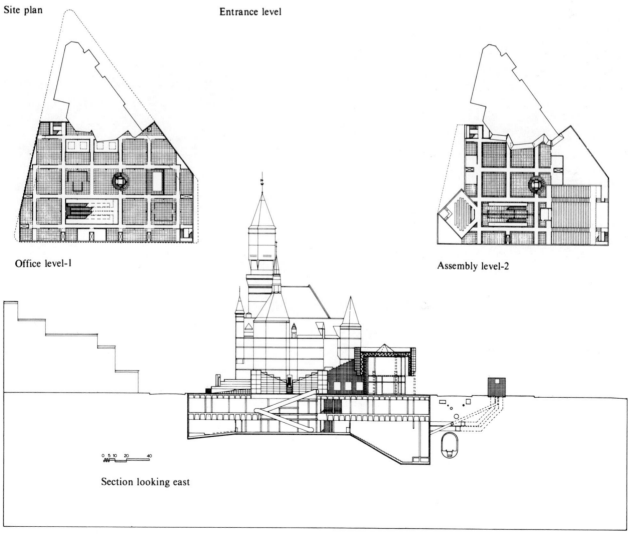

Office level-1

Assembly level-2

Section looking east

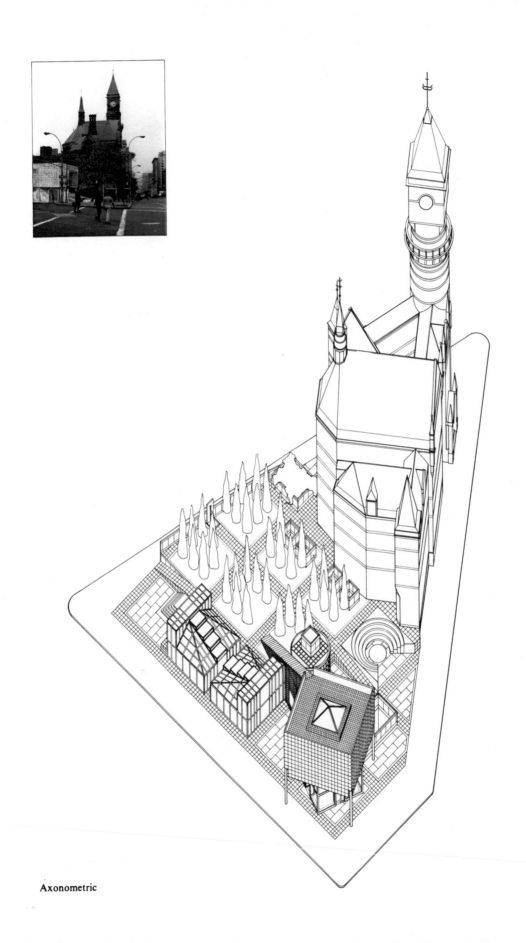

Axonometric

# Research

"The Research category was rather thin in numbers. In addition, the jury had trouble judging them because there was such a disparity in the kinds of questions that were being addressed. It was difficult to compare one project to another; and also, it seems that many people sent in projects that were not justifiable as architecture or research. Most were only speculation, but entered as serious research. The only task for the jury was to separate the real research from the pseudo research. Once that was done, all the real research won awards."

Edward Allen

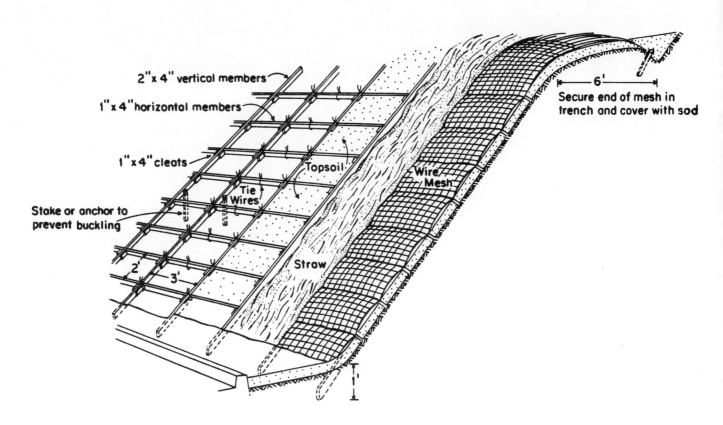

2"x 4" vertical members

1"x 4" horizontal members

1"x 4" cleats

Stake or anchor to prevent buckling

Tie Wires

Topsoil

Straw

Wire Mesh

6'

Secure end of mesh in trench and cover with sod

2'   3'

1'

# First Award

## A Regional Analysis of Ground and Above Ground Climate
## by Kenneth Labs, Undercurrent Design Research, New Haven, Connecticut

This submission is a synopsis of a report prepared for a federal agency on the regional suitability of earth covered roofs and earth-sheltered walls, both on their own terms and in comparison to other accepted techniques of passive design.

The first section, Climatography through References, reviews the state of the art of describing climatic zones for design and the literature of ground climate related to building design. Also contained here is a detailed discussion of models of undisturbed ground temperature and heat flow and heat storage in the soil. Applications are discussed with reference to heat storage in the crawl space of an envelope house and the relationship between undisturbed ground temperature and observed temperatures surrounding two existing earth-sheltered structures. The premises behind the bioclimatic method of weather analysis also are presented.

This section has the dual purposes of establishing the method used for obtaining data for the analysis section, and identifying the literature and issues being analyzed for future workers in the field. It is expected that this section will be of interest primarily to researchers and serious students of ground climatography.

The second section, Regional Suitability of Earth Tempering Practices in Architectural Design, discussed the advantages and disadvantages of earth tempering in different climates, with particular emphasis on whether (and what nature of) earth tempering is compatible with other appropriate climate control strategies on a region-to-region basis. The potential for ground cooling, the likelihood of condensation, the value of recessed placement and earth covered roofs are among the issues considered. The section concludes with a map summarizing the principal findings of the study in a schematic way.

The second section presents the design-related conclusions of the entire report. Data is given with discussion of each issue, but the section is essentially nontechnical. It is written for an audience of consumers, public and private agency officials, professionals and paraprofessionals. It is intended for possible distribution as a separate "pull out" section of the parent report or to be reproduced in whole or in part in other publications to aid in its dissemination.

The third section, Regional Variation in Strategies of Climate Control, presents the results of the bioclimatic analysis of long term weather data performed for this report. It also includes published solar design data to provide a succinct portrayal of local climate for the designer. This compilation should be a useful resource for both architects and engineers at several levels of design development. In terms of this report, the data is offered to those who wish to challenge or examine in further detail the conclusions made in the regional suitability section.

The fourth section, Ground Temperature Tables, consists of a set of tables which provides for the first time a ground temperature data base for the contiguous United States. Although the use of synthesized predictions of ground temperature has limitations, the absence of any useful body of ground temperature data should make this reference section of considerable interest to engineers, designers, experimenters, and other disciplines wholly unrelated to the building industry. The tables are of particular value to those investigating the use of earth-air heat exchangers ("earth pipes"). The tables are included in the report as an appendix that could be distributed independent of the parent report.

The conclusions made here are intended to discuss trends and the broadest implications of the climatic analysis. They are not intended to be read as the final word on the suitability of earth tempering practices. The hope is that the "first cut" conclusions will stimulate interest in discussion and a close examination of the raw data on the part of the reader.

This project delineates the present state of the art of analysis concerning regional suitability of earth tempering practices, as well as making the important new contributions of 1) a synthesized ground temperature data base in the form of easy to use tables; 2) the first attempt to assess the initial potential of ground coupling for cooling; 3) a bioclimatic analysis of weather data for 30 cities in the U.S. with reference to the applicability of ventilation, thermal mass, dehumidification, evaporative cooling and other strategies for climate control; and 4) the first documented effort to assess the role of earth tempering within the broader context of other passive design options.

# Second Award

## Biotechnical Earth Support Systems
by Prof. Donald Gray, University of Michigan, Ann Arbor, Michigan

The purpose of biotechnical earth support systems is to provide both support and landscaping on vertical or sloping earth surfaces and to enhance the versatility and appeal of earth-sheltered construction.

Research on the mechanics of fiber reinforced earth is being carried out in order to understand and to quantify the contribution of plant roots and other fibrous reinforcement (e.g., geotextiles) to the shear strength and stability of soils. Simple theoretical models of a fiber reinforced soil (simulating a soil mass permeated by roots) have been developed.

Laboratory direct shear tests of soils reinforced with different types of natural fibers, e.g., palmyra fibers, reeds and broom straw, are being conducted to test the validity of the models and to determine the influence of fiber-soil parameters such as: fiber orientation, length and diameter of fibers, tensile strength and modulus of elasticity, skin friction of fibers, and concentration or area ratio of fibers. Some typical results of direct shear tests on fiber reinforced sands are included.

The principle of biotechnical earth support systems is based on the fact that both plants and structures function together in a complementary manner to retain earth fills and protect earth slopes. Structural elements (cribbing, lattice work, ties, stakes, and facings) provide mechanical stability and resistance to earth forces. Structural facing must be porous so that vegetation can be planted and established in the interstices.

Vegetation provides erosion protection and some mechanical restraint via root reinforcement. Plants also help to stabilize the ground by extracting and transpiring moisture.

Biotechnical earth support systems utilize low cost, native materials and tend to be less "energy-capital" intensive than conventional structural support systems.

Plants reduce heating and cooling loads in adjacent buildings by moderating ground temperatures via transpiration, shading and wind deflection.

Biotechnical earth support systems fall within a spectrum of approaches which range from "live construction" methods (e.g., seeding and sodding) on the one hand to "inert construction" (e.g., reinforced concrete retaining walls) on the other. Biotechnical or "mixed construction" methods include: vegetated, cellular revetments; vegetated, open-face crib walls; vegetated, wire mesh reinforced earth; contour wattling and brush layering. Examples of some of these biotechnical methods are illustrated.

Biotechnical methods can be used to stabilize, support, and landscape earth berms and embankments, cuts and recesses in slopes, exterior, earth or rock-fill, dividing walls, and tiered or benched slopes and excavations.

This research project was supported by the National Science Foundation, NSF Research Grant No. CME 7910553.

# Second Award

## Evaluation of Free-Span Earth-Sheltered Structure and its Method of Production
by Richard Peterson, Student, University of Minnesota, Minneapolis, Minnesota

The goal of this research proposal is to test a means of producing earth-sheltered structures which allow free-span interiors, modular construction, above as well as below grade development, passive solar configurations at costs competitive with conventional dwellings.

Preliminary research indicates that a modular, fiberglass forming system, designed to produce thin-shell, steel-reinforced concrete structures, is a means of achieving this goal. This research involved constructing a small scale prototype forming system, pouring a shell and live-load testing the finished structure.

The small (20 foot diameter) concrete shell was load tested in four increments of 6,250 pounds, resulting in a total load of 25,000 pounds. This is in accordance with "Building Code Requirements for Reinforced Concrete" as published by the ACI 318-77 (Chapter 20). This test was done under the supervision of Dr. Ladislav Cerny, President and Associate Professor of Structural Engineering at the University of Minnesota, Department of Civil Engineering. As concluded in the final test report:

"The tested shell showed very small deflections andpractically complete recovery after the load removal. During the test, no cracks or any other signs of distress were observed . . . Thus it is concluded that the shell passed the test successfully and is safe for the service load of 50 lbs/ft$^2$."

Though the forming system was successful, significant improvements were suggested by the field test. This proposal is directed at the eventual full scale field testing of an improved forming system. The improved forming system would be more durable, easier to erect and capable of withstanding the load pressures of monolithic pours.

The most significant aspect of the proposed research is that it is not just concerned with end-product architectural considerations but with engineering and production considerations as well. The proposed forming system would be labor efficient and materials efficient with high structural integrity. In addition, it could be assembled into numerous configurations allowing for versatility of design and energy efficiency.

The decision-making process that resulted in this design was also interesting. The first step was the construction of a small (16 foot diameter) underground geodesic dome. The thermal properties of this structure were excellent: cool on very hot days, warm in winter. Thus the benefits of building underground were confirmed. This led to a series of drawings, scale models and another actual-scale structure. It became clear that a low-cost means of production was necessary for such buildings to be feasible on the open market, so a prototype forming system was designed, constructed and used to pour the actual-scale structure described above. This led to the development of another series of drawings and scale models, which resulted in a design for full-scale single-family dwelling.

With earth-sheltering and good solar design a house produced with this forming system would be very energy efficient. To demonstrate how this proposed building would perform during a Minnesota winter, a computer simulation of the building's thermodynamic properties during the winter of 1976-77 was run. The program was written in Fortran IV. The computer used was the PDP-1170. The program takes into account the sun's position hourly through the day, its seasonal declination, the water heating collector's latitude and orientation, the size of the collector and glazing area, sky conditions (cloudiness factor), outside temperature, a steady wind, thermal losses through the building envelope and glazings, and the thermal mass of the building. Weather data is from the U.S. Weather Service at the Minneapolis/St. Paul International Airport. The results of this program indicate the need for 10,300 kBTUs for the winter or 1,600 BTUs per degree day.

In summary, this research project includes in its scope not only the energy efficiency of the design (see energy analysis), but also the structural efficiency (see prototype load testing), the materials efficiency (thin shell, but also adding significant thermal mass) and testing the means of production (the forming system itself).